Spaces of Work: Global Capitalism and the Geographies of Labour

Spaces of Work: Global Capitalism and the Geographies of Labour

Noel Castree, Neil M. Coe, Kevin Ward
and Michael Samers

SAGE Publications
London • Thousand Oaks • New Delhi

First published 2004

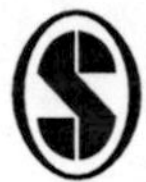

SAGE Publications Ltd
6 Bonhill Street
London EC2A 4PU

SAGE Publications Inc.
2455 Teller Road
Thousand Oaks, California 91320

SAGE Publications India Pvt Ltd
B-42, Panchsheel Enclave
Post Box 4109
New Delhi 100 017

British Library Cataloguing in Publication data

A catalogue record for this book is available from the British Library

ISBN-13: 978-0-7619-7217-4

Library of Congress Control Number 2003102340

Typeset by C&M Digitals (P) Ltd, Chennai, India
Printed and bound in Great Britain by Athenaeum Press, Gateshead

Contents

Acknowledgements

This book has its origins in a third year undergraduate course entitled 'Globalization, labour and locality' co-taught by Noel Castree, Richard Meegan and Michael Samers at the University of Liverpool from 1995–2000. When Noel moved to the University of Manchester in 2000 the intention was still that the trio write what was to have been called *Globalization and the Geographies of Labour*. However, Richard subsequently proved too busy to take on any more writing commitments, at which point Noel's new Manchester colleagues, Neil Coe and Kevin Ward, kindly and enthusiastically stepped into the breach. So, if the book began its life in Liverpool, it ended it bearing the intellectual imprint of the other city in the North West of England, Manchester! Though the present book is 'Meeganless', the four of us nonetheless owe Richard a debt of gratitude for his ideas and encouragement. We are also indebted to the various people who reviewed the synopsis of the book – especially Dick Walker. These reviewers made very constructive suggestions that have made the book better than it would otherwise have been. We also wish to thank the three manuscript readers. We asked two of our postgraduate students (Katie May and Katie Morrow) to read draft chapters with an eye to their intelligibility. The candid feedback they gave us was very useful and much appreciated. David Sadler, whose student days are well behind him(!), kindly agreed to serve as a peer reviewer. The numerous insightful comments and criticisms he made have been a big help and we thank him. Nick Scarle, the School of Geography cartographer, is responsible for producing the many figures and diagrams in the book. We're indebted to him for his sterling technical skills.

Though this book has been a collective endeavour, each of us has taken responsibility for leading on the writing individual chapters and for editing others. Noel took the lead on Chapters 1 and 9, Neil and Noel on Part One, Neil and Kevin on Part Two, Noel and Michael on Part Three. Though we all accept equal responsibility for the overall product, we're also willing to take individual blame if certain chapters are found wanting! Finally, we'd like to thank our publisher, SAGE, the production team (Seth Edwards, Vanessa Harwood and David Mainwaring) and particularly Robert Rojek. Like so many writing projects, this one took a very long time to come to fruition. Robert's patience has been much appreciated.

Finally, every effort has been made to seek permission from other publishers to reproduce the tables, figures, maps and diagrammes used in this book. The authors would like to acknowledge Blackwell Publishing Limited, SAGE Publications Limited and Thomson Publishing Services for permission to reproduce material published in their lists.

List of acronyms used in the book

AFGWU	American Flint Glass Workers' Union
AFL-CIO	American Federation of Labour-Congress of Industrial Organizations
AFSCME	American Federation State County and Municipal Employees
AHC	Anchor Hocking Corporation
AID	Association for Industrial Development
ALGUS	Alliance and Leicester Group Union of Staff
ASEAN	Association of Southeast Asian Nations
BLW	Baltimore Living Wage
BUILD	Baltimoreans United in Leadership Development
CBI	Confederation of British Industry
CEO	Chief Executive Officer
CSSA	Computing Services and Software Association
EP	European Parliament
EWC	European Works Council
FDI	Foreign Direct Investment
GATT	General Agreement on Tariffs and Trade
GCC	Global Commodity Chain
GDLP	Global Division of Labour and Power
GDP	Gross Domestic Product
HDD	Hard Disk Drives
HRM	Human Resource Management
IMF	International Monetary Fund
ICEM	International Federation of Chemical, Energy, Mine and General Workers' Unions
ICFTU	International Confederation of Free Trade Unions
IDL	International Division of Labour
ILB	International Brotherhoods of Longshoremen
ILO	International Labour Organisation
ITS	International Trade Secretariat
JIT	Just-In-Time

LETS	Local Exchange Trading Systems
LLCR	Local Labour Control Regime
MCC	Mondragon Cooperative Corporation
MDHC	Merseyside Docks and Harbour Company
NAFTA	North American Free Trade Agreement
NCL	National Consumers' League
NIE	Newly Industrialising Economy
NIDL	New International Division of Labour
OECD	Organisation for Economic Cooperation and Development
PESO	Philippine Employment Service Office
PEZA	Philippine Economic Zones Authority
RAC	Ravenswood Aluminium Corporation
R&D	Research and Development
SIGTUR	Southern Initiative on Globalisation and Trade Union Rights
SME	Small and Medium Size Enterprise
SSC	Stop Sweatshops Campaign
TCC	Transnational Capitalist Class
TELCO	The East London Communities Organisation
TNC	Transnational Corporation
TUC	Trades Union Congress
UAW	United Auto Workers
UDC	Urban Development Corporation
UNCTAD	United Nations Conference on Trade and Development
UNITE	Union of Needletrades, Industrial and Textile Employers
USWA	United Steelworkers of America
WCL	World Confederation of Labour
WFTU	World Federation of Trade Unions
WITSA	World Information Technology and Services Alliance
WTO	World Trade Organisation
WWW	Women Working Worldwide

Preface: the Landscape of Labour

> It is of the utmost importance to stress that we live in a world in which capitalist social relations *are* dominant, the rationale for production *is* profit, class and class inequalities *do* remain, and that wealth distribution *does* matter. Ray Hudson, *Producing Places* (2001: 2)

Three billion individuals on this planet are wage labourers. They are employed by only 50 million business men and women worldwide. The wealthiest of these business people, like Bill Gates and Rupert Murdoch, command assets equivalent to those of several small countries. By contrast, the people who work for them each command a miniscule fraction of the world's wealth. What's more, they must labour for some 65–80 per cent of their lives if they are to satisfy their everyday needs and wants. Their salaries help to sustain not just themselves but entire families and communities. These wage-workers are young and old, unskilled and skilled, male and female, able bodied and disabled, gay and straight, educated and uneducated, single and married, cohabitees and divorcees. They work in every conceivable economic sector, from farming to manufacturing to services. They exist in every place on this planet, from the most remote Chilean village to mighty cities like Los Angeles and Tokyo. Their pay and conditions of work vary enormously both within and between places, but they have one fundamental thing in common: they sell their capacity to work – their labour power – in return for money. Meanwhile, a staggering 160 million people – a number almost equivalent to the population of Brazil – are currently seeking paid employment. Furthermore, if this vast reserve army of the unemployed is not to grow even larger, some 500 million new paid jobs will have to be created worldwide by 2010. Without these new jobs, population growth will produce unprecedented levels of global, national and local unemployment (ILO, 2001). Finally, as if this were not enough, tens of millions of people are today uprooting themselves from their homeplaces and criss-crossing the globe in search of employment. Labour migration on this scale has not been seen since the turn of the last century and is unlikely to slow down in the near future.

This book is about the more than half of humanity that works – or is seeking to work – for a wage. It is written for degree students (and their teachers) in the social sciences taking courses on labour in the modern world. If not now, then in the future, this book is thereby about you the reader – just as it about us the authors: for students, like academics, must earn a living and they tend to do so by labouring for others in return for money. Indeed, so 'normal' has it become in the modern world for most people to offer themselves as wage-workers that it's easy to forget what a relatively new norm this is. One can work for oneself; one can work for others but for non-monetary returns; and one can work with others in order to produce things that are not exchangeable for coins and notes. In some parts of the world – especially in the South – these forms of non-waged work persist, albeit as islands within a vast sea of paid labour. But the overwhelming reality of work in the twenty-first century is that it entails, in effect, 'selling oneself' to an employer for certain daily, weekly, monthly and yearly periods. Most of this paid labour goes on in workplaces separate from the home, though in a minority of cases the latter remains the main site of commodity production.

Another way of saying this is that we live in a distinctively *capitalist* world. Though capitalism is by no means new, as a way of producing goods and services it is, today, globally dominant. In capitalist economies, businesses large, medium and small make things with the overriding aim of making money – and more particularly profit. The logic of capitalism is thus not about, say, social equality, human happiness or environmental justice. These laudable things may now and then be a means to the end of making money or even its outcome, but in capitalist systems they are rarely ends in themselves. In this context workers are, in essence, one 'factor of production': their services are purchased by employers along with other 'inputs' in order to make commodities that can be distributed and then sold to consumers. According to Marxist and neo-Marxist theorists, this means that whatever their other differences wage labourers together comprise a 'working class'. Globally, this working class is larger than at any point in human history. Certainly, it vastly outnumbers the 'capitalist class' that employs it in different workplaces worldwide. So why, it might be asked, do so many wage labourers worldwide work for a pittance in appalling conditions (be it immigrant female garment workers in New York or bonded child farm labourers in Pakistan)? Why is it that others, even when they enjoy better pay and conditions than the worst-off workers, can lose their jobs overnight? Why is it that some workers have to migrate afar simply in order to earn a living wage? And why do paid workers worldwide not unite in a mass movement to improve the circumstances under which *all* of them labour?

These are fundamental questions: they are questions about labour's present condition and future prospects in the contemporary capitalist world. And they are questions we seek to answer in this book. In so doing we want to draw readers' attention to the profound importance of geography – or rather geographies, in the plural. Labourers do not live and work on a global isotropic plain nor on the head of a pin. Rather, they are located in a landscape of geographical difference and geographical interconnectivity. Workers always live and work somewhere – even migrant workers. As Ray Hudson (2001: 122) puts it, 'Labour is the most place-based of the factors of production'. So if we are to understand what is happening to workers today we must attend to the importance of *place*. But equally, in this era when capitalism is an increasingly global phenomenon, we need to appreciate that what happens to workers in one place is incomprehensible without paying heed to inter-relations extending across *space*. Workers, and the businesses employing them in specific places, are more than ever connected to distant others within a national, international and global space economy. As one commentator argues, today 'local happenings are shaped by events occurring many miles away, and vice versa' (Giddens, 1990: 64). By virtue of these increasingly intense translocal connections workers in different parts of the world may be pitched into relations of competition or cooperation depending on the circumstances. To understand which of these relations comes to the fore in particular cases we need, as this book will explain, to appreciate the importance of *geographical scale*. Scale is, if you like, the middle term between place and space. It concerns whether and how local (or sub-local) scale events and actions reverberate across space (and vice versa). The capacity to 'up-scale' actions from a place or places to larger spatial scales can be an enormous source of power for particular businesses or workers. For instance, transnational companies have the capacity to search the globe for suitable locations for certain of their production facilities. Likewise, the capacity to 'contain' actions – like a workers' strike – within a certain scale (such as the local) can be a powerful weapon for (or against) employers and labourers in pursuit of their respective objectives.

So place, space and geographical scale are the watchwords of this book. Triangulated, they allow us to map the complex landscape in which contemporary workers live, work and struggle. It is, as we shall see, a landscape riven with elemental paradoxes and dilemmas. These are profoundly *geographical* paradoxes and dilemmas. To analyse labour in a non-geographical way is therefore, we submit, to miss much of what is fundamental to the everyday lives of workers in this era of global capitalism. Geography must be present in the analysis from the very start: for place, space and scale, we shall argue, both profoundly structure and are profoundly structured by the practices of workers, as well as the practices of the other major social groups and institutions the fortunes of workers are intimately tied to. By insisting on the centrality of our geographical triad to understanding the current condition and future prospects of workers worldwide we are saying that these workers – and those who analyse them – should equip themselves with a *geographical imagination.*

Of course, to assert that 'geography matters' in this way has become a familiar refrain in the discipline of human geography. But it's also resonated in other fields like sociology, economics and cultural studies in the last few years, where such luminaries as Anthony Giddens, Paul Krugman and Fredric Jameson have embraced the 'geographical turn'. Accordingly, this book is written for students and teachers of labour issues across the social sciences. It is not, in other words, a narrowly disciplinary book that seeks to police boundaries between 'geographical' and 'non-geographical' approaches to labour. For we wish to collapse this distinction by making geography a necessary, but not privileged or sufficient, part of any proper understanding of labour in the modern world.

All this said, it may seem that our writing this book has come too late. After all, in the few short years that have passed between conceiving of and actually completing this volume, a plethora of texts on labour under conditions of global interconnectivity has been published. In geography, following on the heels of Jamie Peck's (1996) *Workplace,* Andrew Herod has blazed a trail with his (respectively) edited and authored books *Organizing the Landscape* (1998) and *Labor Geographies* (2001), added to which is *Space, Place and the New Labour Internationalisms* (2001), co-edited by Jane Wills. Beyond this quartet of geographically full-blooded labour studies have been others with a more implicit geographic sensibility, like *Working Classes, Global Realities* (Panitch and Leys, 2001), *Labour Worldwide in an Era of Globalization* (Munck and Waterman, 1999), *Globalization and Patterns of Labour Resistance* (Waddington, 1999), *The Global Economy, National States and the Regulation of Labour* (Edwards and Elger, 1999), *Globalization, Social Movements and the New Internationalisms* (Waterman, 1998) and *Workers in a Lean World* (Moody, 1997). Yet despite the existence of these several 'labour in an interdependent world' studies, it seems to us that neither singly nor together do they deliver what we hope to deliver in the following chapters. This is so for three reasons.

First, none of these books are written for degree students or their teachers. They are largely research level contributions. Yet many universities now offer undergraduate and postgraduate modules on the subject of labour's prospects in the twenty-first century. This book seeks to distil the ideas and insights of the fast-growing research literature on this subject and present them in an accessible way. It is not an extended review of others' research. Rather, it uses this research (and our own) to explain and illustrate what we think are the key issues. It is pitched at final year undergraduates and postgraduate students in the social sciences, and those who teach them. For too long students studying economic geography, economic sociology, international political economy and the like have been taught a lot about production, trade and finance but virtually nothing about workers. In our own discipline Peter Dicken's (2003) excellent *Global Shift* – a book about transnational companies (TNCs) – is now in its fourth edition. Meanwhile, virtually no student-focussed texts exist about the tens of millions of wage-workers TNCs directly employ, not to mention the mass of labourers employed in medium and small business (SMEs) worldwide. *Spaces of Work* will, we hope, fill this conspicuous gap in the textbook coverage of workers in an interdependent world. Because we want this book to be highly accessible for students we have self-consciously sought to avoid some traps that, in our view, too many textbooks fall into: namely, cluttering the text with references, rehearsing arcane theoretical debates, and writing in dry, dispassionate, third-person prose. We have also provided carefully selected further readings accompanied by key questions, as well as a glossary and list of relevant websites (these can be found at the end of the book).

Second, there is an issue of theoretical synthesis. The new books on local and global labour cited above tend to treat different aspects of labour's existence in isolation. Thus Peck is very good at theorizing local labour markets but says next to nothing about worker resistance. Likewise Herod's *Labour Geographies*, while excellent on the twists-and-turns of employer–worker struggle, virtually ignores social reproduction, migration and non-union labour organizing. In *Spaces of Work, Geographies of Labour* we pull these theoretical fragments together to provide a comprehensive conceptual map. Finally, there is a question of empirical coverage. There are some notable biases and blindspots in the research literature on workers in the contemporary world. One of these is an undue obsession with manufacturing, as if this were the only or most important economic sector globally. Another is a tendency to focus on Western workers who, by comparison with most of those in the South or the former communist bloc, are relatively well off. In this book we seek to redress this sectoral and geographic imbalance by presenting a variety of examples and case studies. As we observed above, wage labour exists anywhere and everywhere today and it is thus incumbent on analysts to present something of this promiscuous diversity.

So far so good, but what about the politics of *Spaces of Work*? By politics we mean our ethical stance towards wage-workers. We ask this question because it would be dishonest to imply that we take a neutral perspective on labour in this book. All four of us, in our research publications, have sought to apply and develop Marxist, neo-Marxist and institutionalist perspectives on the geography of wage work/ers (see Appendix 1 for an explanation of these perspectives). Consequently, we align ourselves with that broader movement called 'critical social science'. This puts us on the Left of the political spectrum. As the revolutionary Karl Marx once famously said, the point is not simply to understand the world but to change it. To be 'critical' is to offer a negative judgement on something. To 'criticize' is a political act because it calls some aspect of our current world into question and thereby implies a set of preferable future arrangements. As such, criticism is always launched from specific 'normative standpoints' (Sayer, 1995): that is, value systems held by critics that underpin their assessment of the current state of affairs and their recommendations for desirable future changes. Working in a capitalist world can be a brutal affair, for both workers and their dependents. Our normative standpoint is broadly Marxist in two senses. First, we believe that while wage-workers produce much of the world's wealth they receive very little of it – and that this is wrong. Second, we believe that all wage-workers should be entitled to basic rights and rewards, regardless of place, 'race', creed or colour. When and where these elemental entitlements are not enjoyed we argue that a social injustice has been done. As we shall see, labourers in a capitalist world routinely suffer all manner of basic work-related injustices.

However, over and above basic rights and rewards the question of what is 'good' and 'bad' for workers gets more complicated. And much of the complication is attributable to geography. As we will argue, what happens to workers, and what workers do in response, is ineluctably *context specific* – that is, conditional on how place and space (which is to say, geographical scales) articulate in particular cases. It therefore follows that one 'cannot make absolute determinations' (Herod, 2000b: 1788) about the current conditions and appropriate future actions of specific workers in specific localities. The corollary is that it is misguided to offer criticism of diverse cases on the basis of universal, rigidly held normative standpoints, just as it is inappropriate to make generalizations about the likely fate of workers in the modern world. And yet several of the authors we have mentioned above – all of whom are critical social scientists – have at various points given in to the temptation to make sweeping judgements about labour. Peck, for example, ends *Workplace* with an apocalyptic vision of hypermobile multinationals playing workers in different places off against one another in a 'war for jobs and investment'. Likewise, Munck and Waterman's (1999: x) opening gambit in *Labour Worldwide in the Era of Globalization* is that workers are 'undoubtedly in difficult straits'. Such blanket judgements are, in our view, unsupportable. They tend to be based on limited evidence and they often presume that a universal standard of what's 'good' and 'bad' for workers can be applied worldwide. In the penultimate chapter we confront this problem of how one judges the in/justice of workers' circumstances head-on by arguing that geographical scale is the key. Determining what is just and unjust for workers must depend on the scale – sub-local, local or supralocal? – at which justice is being sought.

Knowledgeable readers will have noticed that we've made no mention of the classic Marxist idea that wage-workers need to instigate a revolution to topple capitalism. Though in Chapter 8 we consider cases where workers are organizing internationally and globally against capital, the possibilities of replacing global capitalism – and thus wage labour along with related un- and underemployment – strike us as immensely slim. This is partly because of the scale of dilemmas confronting workers, which we explore in Chapter 9. David Harvey, one of the most uncompromising Marxist critics writing today, has recently pleaded for an 'optimism of the intellect' (2000: 17), even when circumstances seem unpropitious for progressive change. Though sympathetic to Harvey's injunction not to cave in to reformism, we feel that many workers have considerable 'wiggle room' within the confines of capitalism to improve their situation. Using this wiggle room to an advantage is a fundamentally geographical project. In Chapters 6–8 we pinpoint the different elements of this wiggle room and argue that agitating within – rather than against – global capitalism does not mean that wage-workers have somehow 'sold out to the enemy'. Even Marx, capitalism's most penetrating critic, didn't see this now dominant system as an unmitigated evil. For millions of workers a more just post-capitalist future might be preferable to a frequently unjust capitalist present. But in the meantime, it's important to locate opportunities for progressive change within the existing system. There is simply too much misery and injustice in this world (as we show in Chapter 5) for labour to pin its hopes on some utopian global project to slay capitalism. Much can be done in the here-and-now.

All this said we do not think that writing a textbook about labour will, in any direct way, assist labour's cause! For years, Left-wing social scientists have been trying to find the means to 'connect' with those disadvantaged, excluded or otherwise marginalized groups they normally study. In light of this, our frank admission that few, if any, of the people this book is about will ever read or learn from it might seem distinctly 'un-Left'. Unless, that is, we remember that there are many roads to political change, some of them less direct than others. In this case, our primary audience is you: students who will one day become (if you are not already) wage-workers and perhaps even represent other employees. If *Spaces of Work* does any political work at all, then, it will be pedagogical. And why not? After all, students are typically the largest audience Left-wing academics ever reach in their lifetimes. If this book gives you pause for thought – if it makes you think hard or think differently about the plight of workers in a capitalist world – then it will have done something useful. Even if, in writing this book, we as authors are merely seeking to understand the world of labour, in so doing we might just help some of our student readers to one day change it.

NCa, NCo, KW, MS
Manchester

A note on geographical terminology

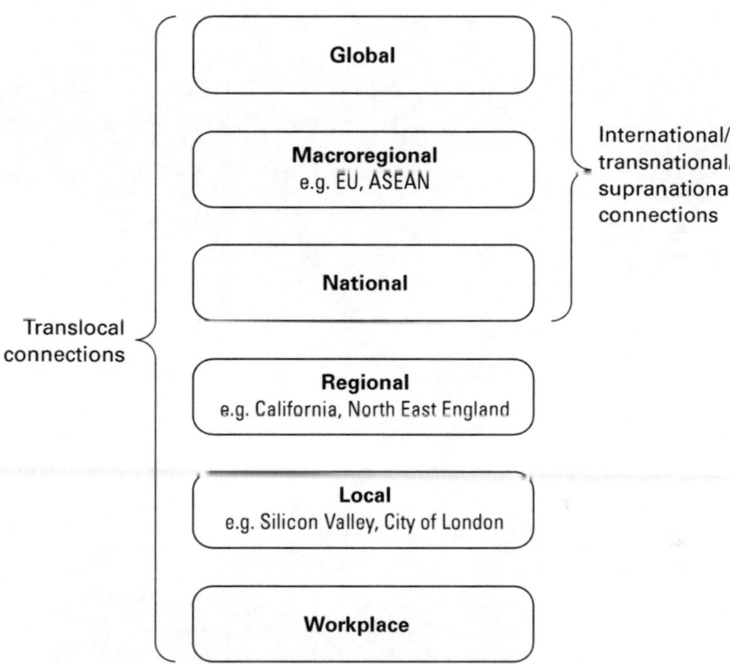

Figure 0.1 Scales of analysis: a geographical vocabulary

In this book we use several geographical terms to describe the spatial scale of events and processes. In order to avoid confusing readers, Figure 0.1 summarizes the vocabulary deployed and indicates the different spatial scales referred to. As the figure indicates, we use the terms 'international', 'transnational' and 'supranational' as synonyms, while using the more general term 'translocal' to encompass relations that link workplaces to any or all larger geographical scales.

one

Orientations

WORKING LIVES: SOME STORIES

This book is about real people who live, learn and labour in real places. It's about lives made, derailed or shattered depending on the capacity of workers to attract, retain or follow opportunities for paid employment. In a very literal sense, it's a book about all of us. Its starting point is the elemental fact that the overwhelming majority of people who work for a living in our world do so in return for a wage. These people, in effect, become commodities for definite periods of daily, weekly, monthly and yearly time. Their capacity to work is 'purchased' by a relatively small number of other people – employers – who own farms, factories, call centres, branch plants, banks, sweat shops, supermarkets and myriad other units of production. Whoever and wherever they are in the world, this mass of wage-workers shares the common experience of labouring for others in order to make a living. In this introductory chapter we introduce the key ideas of this book and set the scene for the chapters to come. We begin with some stories about the working lives this book seeks to make sense of.

Rajan Bhaskar

Rajan Bhaskar is a university educated 24-year-old working in his hometown of Bombay, India. He earns US$3500 a year as a customer service representative in a call centre run by American Express, the US credit card company. By Indian standards Rajan is paid a good salary. Indeed, it's so good that for every job like Rajan's that's advertised over 1,000 Indian university graduates apply. Rajan, like many of these graduates, speaks excellent English, a legacy of India's long colonial ties with Great Britain. He needs to: for virtually all of his clients are calling from the United States. When they ring to alter their credit limit or order a new Amex card they are blissfully unaware that they are talking to someone on the other side of the world. After all Rajan – or Richard, as he must call himself at work – has been trained by his employer to use American phrases (like 'Have a nice day') when he talks to clients. Meanwhile his computer screen shows the current time, weather and major news-stories in the United States – information he can drop into his conversations with clients as if he were speaking to them in mainland America. To employ someone like Rajan in the United States would cost American Express some US$18,000 per annum. As Rajan's supervisor, Raman Roy puts it 'Geography is history. Distance is irrelevant.' Because telephone communication is instantaneous – because it 'compresses' space and time – it is both possible and profitable for American Express to open new call centres in India at the expense of its existing ones in the United States (source: *The Guardian*, 9 March 2001).

Jane McDonald

Jane McDonald is a married, 49-year-old mother of two. Until early 2003 she and her husband were among the 800 UK employees working on an assembly line for one of Britain's newest and most celebrated manufacturing companies: Dyson. Dyson, named after its English inventor-owner James Dyson, is famous for making the world's first 'bagless vacuum cleaner'. This powerful, dual cyclone machine has become the vacuum of choice for countless British households since first marketed in the mid-1990s. In the last two years it's also started to sell well in continental Europe and the United States. Each day, some 8000 'Dysons' leave the company's factory in the small Wiltshire town of Malmesbury (population 5,420). Jane's work on the assembly line is repetitive and dreary, but it pays enough for Jane and her family to get by. Or rather it used to: for the Dyson factory recently closed, leaving Jane, her husband and the other 798 Malmesbury employees jobless. Production is to be relocated to a factory in peri-urban Malaysia, where hourly labour costs average £1.50 per worker per hour as opposed to £4.10. In addition, many of the companies who supply parts for the Dyson cleaners are based in the

Far East. While good news for Malaysian workers, the plant closure has hit Jane's family especially hard because both she and her husband were dependent on Dyson for their livelihoods. Indeed, Dyson was Malmesbury's largest single employer and, with few other jobs in the area, Jane and her husband now face the prospect of either long term unemployment or a lengthy commute to seek work in cities like Bristol (source: *The Guardian*, 6 February 2002a).

Ruben Chacão

Ruben Chacão is just 10 years old but works an average of 11 hours a day, six days a week. He lives with his family in one of the many favelas in and around Rio de Janeiro. He spends his days in a poorly ventilated, ramshackle factory with other under-age workers. Together they stitch together footballs destined for Southern Europe where they'll be embossed with the logos of the national soccer federations of whichever country they're sold in. They'll then be sold to consumers in sports shops for a sum many times higher than it costs to make them. Ruben is paid virtually nothing for his work, just a few cents per football stitched. He enjoys none of the basic rights of legal workers in Brazil. What's more his job is depriving him of an education. Ruben does not want to work for a living but he must. His father borrowed – but was unable to repay – money from the man who is now Ruben's boss. Ruben will thus work until he has paid off his father's debt. He will probably be 13 by the time he's done so. By then he'll be a highly competent stitcher but skilled at little else. Although the governing body of world soccer, FIFA, has been trying to crack down on the use of child labour in football manufacture, it does not have the monitoring systems to identify where many soccer balls are made. At the same time, the Brazilian government – both nationally and locally – lacks the resources to properly enforce its own legislation outlawing under-age work (source: Gerard Oonk, *The Dark Side of Football*, 2001).

Jarrah Ahmed

Jarrah Ahmed is a Saudi Arabian construction worker, born and raised in his country's capital, Riyadh. He's 40 years old and has five children under 15 years of age. During the 1980s and 1990s when Saudi Arabia enjoyed the benefits of sustained oil export revenues, the small building firm Jarrah worked for was almost never idle. However, a late-1990s decline in oil prices combined with an over-supply of commercial and residential property in Riyadh, has meant that building work has dried up in recent times. Desperate for a salary to support his family, Jarrah took up an offer to construct new houses in Gdansk, Poland. This former communist country is fast becoming integrated into the capitalist world economy. Foreign companies have relocated to cities like Gdansk to take advantage of skilled but relatively low cost workers in industries like electronic assembly and mechanical engineering. The consequent demand for new housing by these workers has fuelled a minor construction boom. Because there are simply not enough skilled Polish construction workers to satisfy heightened demand, migrant workers like Jarrah have made up the shortfall. Jarrah earns enough in six months to support his family for a year. His income back home is boosted by a generous child allowance from the Saudi government, which helps pay for clothes, toys and the like. But Jarrah's periodic migration comes at a price. As well as being wrenched from his family and friends, in Poland he lives in a very basic hostel along with other migrant workers of various nationalities. He speaks only basic Polish and, in order to save money, his social life in Poland is very limited. When he does venture out onto the streets some Polish people occasionally scowl at him and mutter things he does not understand. Jarrah's foreman, a Saudi who speaks good Polish, tells him these ordinary Poles are probably angry with him because they believe he and other labour migrants are 'taking their jobs' (source: Samers, unpublished).

Terry Teague, Bruce Robinson and Cees Cortie

Terry, Bruce and Cees are all dock workers. Prior to 1996 they did not know each other. Since then Bruce and Cees have defended Terry's right to keep his job – this despite the fact that all three live thousands of kilometres apart in different countries. Terry is one of 350 dockers who worked for the Merseyside Docks and Harbour Company (MDHC) in the Port of Liverpool, England. Bruce is a docker in Vancouver, Canada, whilst Cees has for 20 years laboured in the Port of Rotterdam, Holland, the world's largest dock complex. What connects the three men? Terry and his Liverpool co-workers were fired by the MDHC in late 1995. By early 1996 he and several other members of the Liverpool dockers' trades union had sent delegations to other ports worldwide. These delegations urged non-British dockers to boycott ships destined for Liverpool in order to put pressure on the MDHC to reinstate the sacked dockers. Bruce and Cees are both union reps in their home-ports, and met the Liverpool delegates when they visited Vancouver and Rotterdam in early 1996. They subsequently urged their fellow Canadian and Dutch dockers to boycott Liverpool-bound ships in coordinated international actions against the MDHC later in the year. In so doing they risked their own livelihoods, because refusing to load ships is a sackable offence. What's more, they travelled to Liverpool in 1997 along with other dockers from around the world to discuss how to further pressure the MDHC into reversing its redundancy decision. Though they may never meet again, Bruce and Cees have shown remarkable loyalty to Terry and his workmates living so many miles away (source: interviews conducted for Castree, 2000).

SIMILARITIES AND DIFFERENCES

What do these stories tell us? The most obvious answer is that the countless numbers of people who sell their labour in the contemporary world are astoundingly *diverse*. In our five stories – a tiny subset of those we could have told – we have an assembly workers, dockers, a builder, a tele-service rep and a soccer ball stitcher; we have men and women, adults and children; we have the employed and the unemployed; we have those compelled to work intolerable hours for little reward and those labouring for a good salary in a congenial environment; we have migrants and non-migrants; we have unionized and non-unionized workers; we have a plethora of nationalities, cultures and languages spanning the so-called 'first', 'second' and 'third' worlds; and we have skilled, semi-skilled and unskilled workers. These differences are irreducible. Together they comprise what is known as the *social division of labour* (Figure 1.1). There are three principal ways of categorizing this social division. One way is to distinguish production, circulation, consumption and reproduction workers (Sayer and Walker, 1992: Chapter 2). The first of these makes goods and services, the second moves them to markets, the third sells them, and the fourth maintains them. Alternatively, we can distinguish formal/informal, temporary/permanent, home/non-home and legal/illegal wage-work, as shown in Figure 1.2. Finally, as Figure 1.3 indicates, we can distinguish public from private sector workers and, within the latter, those labouring in local, national or transnational firms. These three taxonomies are not mutually exclusive. Added together, they convey something of the complexity of the modern social division of labour.

Confronted with this befuddling diversity, it may seem hard to identify what contemporary wage workers have in common – other

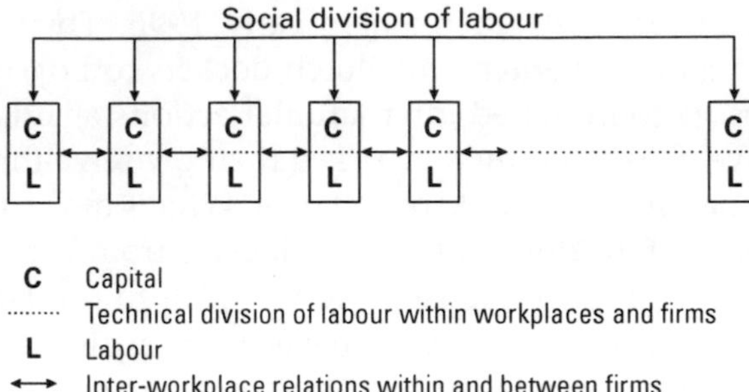

Figure 1.1 The social division of labour (adapted from Sayer, 1995: figure 3.1)

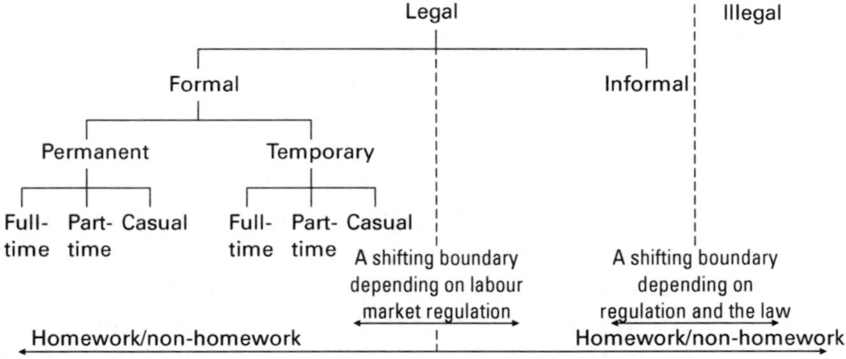

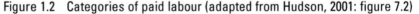

Figure 1.2 Categories of paid labour (adapted from Hudson, 2001: figure 7.2)

than the already-stated fact that they all labour for money. But this fact hardly seems a sturdy enough basis on which to argue that understanding the present and future of paid labourers' daily existence worldwide can be brought within a common frame of analysis. We agree: it isn't. But, if we stand back from the evident differences between and among workers then other deep-seated similarities start to become apparent.

These are simultaneously *social* and *geographical*. Socially, the fact of selling one's labour power has a number of important implications for all workers. For instance, hours worked and wages earned (assuming a person is employed) have a direct impact on the reproduction of workers and their dependents. Reproduction is the daily process of feeding, clothing, sheltering and socializing people. As Peck (1996: 39) observes, production and reproduction 'are both separated and

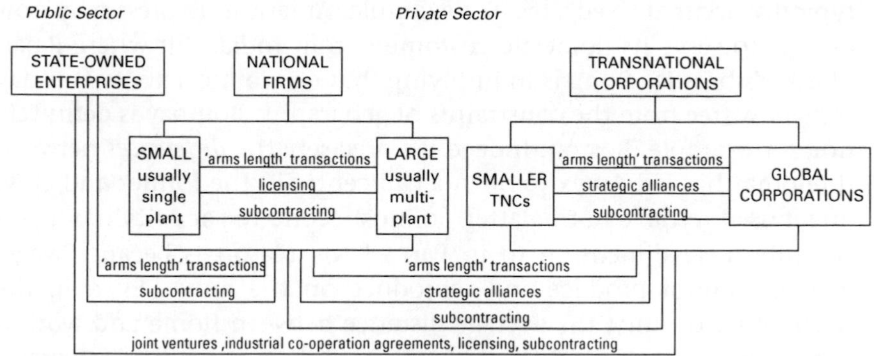

Figure 1.3 A typology of organisation employing wage labour (reprinted from Dicken, 1998: figure 1.3)

connected'. Reproduction occurs principally in the home and the community. Without adequate non-work time and sufficient work-time renumeration, paid labourers of all stripes cannot sustain basic physical reproduction (that is the provision of sufficient food, shelter, clothing and warmth) or social reproduction (making friends, enjoying leisure pursuits or receiving formal education or training, for example). Equally, when pay and conditions are good these two forms of reproduction can be easily and pleasurably undertaken. This is especially the case when the local and national state directly and substantially supports reproduction, as in Jarrah Ahmed's case. But equally, when wage-workers suffer poor pay and conditions or when they are unemployed (in the short term or long term), the state can have an important role to play in ensuring their physical and psychological survival. In the case of Jane McDonald and her husband, the 'jobseeker's allowance' provided by the British 'New Labour' government will take some of the sting out of their job-loss, though Ruben Chacão, alas, will receive no relief from an over-stretched Brazilian state.

The link between paid work (production) and reproduction is just one of several social similarities in the situation of workers worldwide. But, as our five stories illustrate, we cannot understand these similarities in a non-geographical way. There are three key things to say here. First, all the workers in our stories live and work in place. For all the hoopla about living in a 'global village', the fact remains that all production (and indeed reproduction) has to occur somewhere: it remains deeply *place-based*. This is an intractable geographic necessity. The capitalist production of commodities – be they Dyson vacuum cleaners or soccer balls or new buildings or credit card services – must typically occur at fixed sites. For example, American Express may now be able to serve its domestic customers from India. But where Rajan Bhaskar's boss is wrong is in implying that economic activity is today somehow free from the constraints of geography. Bombay is definitely not, for example, Boston. Indeed it's precisely the *differences* between them that has led Amex to open a call centre in the former and close one it used to have in the latter. So most economic activity has a persistently local dimension to it. Part of the reason is because wage-workers cannot produce and reproduce on the move. Even in the twenty-first century, the average distance between home and work is but a few kilometres. Though migrant workers exceed this average, even they must still work somewhere – albeit distant from their home-place. So the reality of paid work, even in our shrinking globe, is one of different groups of workers living and labouring in 'local worlds' (Meegan, 1995) that are distant and distinct from one another.

This *spatial division of labour* can be added to the list of social differences between workers (of skill, pay, industry, gender, etc.) mentioned above (we will discuss Doreen Massey's famous analysis of this later in the book). Paradoxically, the fact that workers all *share* a local existence is something that *differentiates them* geographically. But there's more going on than this, which brings us to the second geographical lesson we can learn from our stories. Workers in the twenty-first century are increasingly interdependent. That is, the actions or inactions of groups of workers in one place can have serious implications for other groups of workers in other localities nationally, internationally or globally. To be sure, not all workers are interdependent in the same ways, to the same degree or with the same consequences. But, as our examples show, modern wage-workers have to live with interdependence, like it or not. This 'stretching' of social relationships between place-based workers across a larger national and supranational *space* takes many forms. In the case of Rajan and Jane it's about relationships of inter-place competition producing winners and losers. In Ruben's case, more subtly, it's about workers in Western Europe purchasing soccer balls (for workers are also commodity consumers, let us not forget) that help, inadvertently, to perpetuate child labour in Brazil. In Jarrah's case it's about workers coming from one place to take up jobs in another that is far distant. And in the case of Terry, Bruce and Cees the international relationships forged are ones of cooperation and solidarity. So, in geographical terms, contemporary wage-workers exist in a complex landscape of *geographical difference and geographical interdependence*. This dialectic of place and space produces numerous dilemmas for working people. These are dilemmas of *geographical scale* – the third geographical lesson of our stories. Should workers think and act at the sub-local, local or the translocal scale? Should they accent their differences from workers in other places in order to defend local jobs? Or should they join together with distant others – as our three dockers did – in order to 'up-scale' loyalties and actions? If they do the latter then at what particular scale and with which other workers should they unite? These are momentous geographical questions that are at the heart of what it is to seek and retain paid work in our capitalist world. As we shall see, they admit of no easy answers.

In Part one of this book we want to introduce a theoretical framework that will help us 'map' the social and geographical landscape in which labourers exist in the world of twenty-first-century capitalism. The three chapters that follow this one will be the longest and most conceptual ones in the whole book. They have to be. By the end of them we want readers to have the intellectual tools necessary to understand the basic socio-geographic parameters that both structure and are structured by contemporary paid workers. The subsequent chapters will then add further conceptual layers and much factual detail. Chapter 2 explains the social universe that defines what wage labourers can and cannot do in a capitalist world. The third and fourth chapters then 'add in' geography to show why place, space and scale necessarily enter into the constitution of wage workers' social existence. Together, these three chapters synthesize insights drawn from across the non-geographical and geographical parts of social science. Student readers need not take fright at the prospect of theoretically 'heavy' chapters. Theory often has a bad name among undergraduates and postgraduates. The common complaint is that it's too abstract and divorced from reality. But, as we hope to show in this chapter, abstraction is precisely what allows us to make meaningful statements about the real world. When confronted with a large, dynamic and complex socio-geographical world we cannot, like laboratory scientists, conduct controlled experiments to figure out what's going on. All we can do is look carefully and think very hard in order to get behind the details and thereby identify underlying factors and processes. This is what theorizing in social science is all about. We cut into the connective tissue of the world conceptually, hoping our mental scalpel will allow us to distinguish the key factors. Without theory we are, to use a different metaphor, simply unable to see the wood for the trees. In the following three chapters we proceed by moving from the simple to the complex. In other words, we start with concepts that isolate key aspects of wage-workers' existence, and then progressively

TABLE 1.1 Profiling the global workforce in 2000

Country	Labour force (million)	Labour cost per worker in manufacturing (US$ per year)	Value added per worker in manufacturing (US$ per year)
1 China	756.8	729	2,885
2 India	450.8	1,192	3,118
3 USA	144.7	28,907	81,353
4 Indonesia	101.8	3,054	5,139
5 Brazil	79.7	14,134	61,595
6 Russian Federation	77.7	1,528	n.d.
7 Bangladesh	69.2	671	1,711
8 Japan	68.3	31,687	92,582
9 Pakistan	51.7	n.d.	n.d.
10 Nigeria	50.3	n.d.	n.d.
11 Germany	40.9	33,226	79,616
12 Mexico	40.4	7,607	25,931
13 Vietnam	40.4	n.d.	n.d.
14 Thailand	36.8	3,868	19,946
15 Philippines	31.9	2,450	10,781
16 Turkey	31.3	7,958	32,961
17 UK	29.9	23,843	55,060
18 Ethiopia	27.6	1,596	7,094
19 France	26.7	n.d.	61,019
20 Italy	25.7	34,859	50,760
21 Myanmar	25.4	n.d.	n.d.
22 Ukraine	25.1	n.d.	n.d.

Source: World Bank, 2002, Tables 2.2 and 2.5. (n.d. = no data)

enrich, supplement and complicate these concepts as we go along. As Dicken (2003: x) observes, 'the constraints of language necessitate a linear treatment for something that is not linear at all'. Thus while Chapter 2 begins *analytically*, by the end of Chapter 4 readers should be able to see the landscape of wage-workers *synthetically*. But before we begin our theorization of labour, we need to do three things: first, share with readers some basic factual information about the current condition of wage workers; second, briefly come clean about our theoretical starting point and third, explain our perspective on a phenomenon thought to be closely linked to working people's fortunes, namely globalization.

THE UNEVEN GEOGRAPHY OF WAGE LABOUR

It is worth pausing briefly at this stage to remind ourselves of the sheer scale of the social and spatial inequities that characterize the

TABLE 1.2 Global labour force in 2000, by World Bank income category

Income per capita	Number of countries	Total labour force (millions)	Average per capita income (US$)
Low (<US$756)	64	1,115.1	410
Lower middle (US$756–US$2,995)	55	1,100.4	1,130
Upper middle (US$2,996–US$9,265)	38	288.4	4,640
High (>US$9,265)	25	439.4	27,680
Total	182	2,943.2	5,170

Source: World Bank, 2002, Tables 1.1 and 2.2

contemporary global economy. Perhaps the most obvious question to ask is, where are the bulk of the world's three billion wage labourers located? As Table 1.1 reveals, some three quarters of them (2.23 billion) live and work in just 22 countries, all of which have a labour force of 25 million or more. In fact, almost half of the world's waged workers are to be found in just four countries – China, India, the United States and Indonesia – a simple reflection of the size of these countries' populations. More interestingly, however, Table 1.1 also reveals considerable variations in labour costs and productivity across the 22 countries. In terms of labour costs in manufacturing, rates vary from US$34,859 per year in Italy to just US$671 in Bangladesh. In terms of productivity, value added per worker in manufacturing (that is the monetary value of products over and above the costs of inputs) ranged from US$92,582 per year in Japan down to US$1,711, again in Bangladesh. More tellingly than just numbers of workers, these data start to hint at the range of terms and conditions that wage labourers face around the world.

Ideally, to reveal these different aspects of the uneven geography of wage labour, we would conduct a geographical analysis of a wide range of variables here, including labour costs, labour productivity, skill levels and unionization rates. However, the most comprehensive data available is that on income levels, which allows us to consider inequalities at a range of spatial scales, from the global to the local, thereby mobilizing the geographical terminology that underpins this book (detailed in Figure 0.1). A useful starting point for our analysis is Table 1.2, which breaks down the global labour force according to the World Bank categories of low-, middle- and high-income countries. From Table 1.2, we can see that 75 per cent of the total global waged-workforce of approximately three billion reside in low- or low-middle-income countries, in other words countries where average per capita incomes are less than US$3,000 per year. The third column of the table gives a sense of the income disparities between countries. The average per capita income for the 25 high-income countries – US$27,680 – is no less than 67 times the average across the 64 low-income countries! Per capita income in the richest country, Switzerland (US$38,140) is a massive 293 times that of the poorest, Sierra Leone (US$130), and these gaps continue to widen. Unfortunately, ours is a world in which one in five people (around 1.2 billion) still live on less than US$1 per day.

Figure 1.4 maps income per capita by World Bank category. The geography it reveals can be crudely summarised in three steps. First, there is a heavy concentration of low-income countries in sub-Saharan Africa, and South and Southeast Asia. Indeed, Sub-Saharan Africa and South Asia account for almost 70 per cent of those living on less than US$1 per day (Dicken, 2003). Second, most Latin American, North African, Eastern European and East Asian countries fall into the middle-income categories. Third, the high-income countries are confined to North America, Western Europe and Australasia, with the notable exceptions of Japan, Singapore, Taiwan, Hong Kong and a few Middle Eastern states. We can clearly see, therefore, that there is a *macro-regional* geography to average income levels within the global economy. However, there is also considerable *national* variation in income levels between countries in the broad macro-regions highlighted by Figure 1.4. For example, of the European Union member states in 2000, per capita incomes varied from US$32,380 in Denmark to US$11,120 in Portugal, a disparity that will be exacerbated considerably by the planned accession of a range of Eastern and Central European countries.

Income is also extremely unevenly distributed *intra-nationally,* as Table 1.3 reveals for a selection of leading economies. We can see that it is not unusual for the richest 20 per cent of the population to receive over 40 per cent of national income, while the poorest 20 per cent receive considerably less than 10. Even in the most equitable societies – for example Sweden, France and Japan – the richest fifth of the population receive a third of national income, and the poorest fifth barely 10 per cent. In some newly industrializing countries – such as Malaysia and Mexico – the inequality is much worse, with the top fifth receiving over half of national income, and the bottom fifth less than 2 per cent. In 1998 Brazil was one of the most unequal nations of all, with the top fifth receiving almost two thirds of total income, leaving the poorest fifth with just over 2 per cent. As Figure 1.5 illustrates for the case of *regions* of the UK, these intra-national inequalities have a clear spatial dimension. The map shows how, in 1999, regional GDP per capita varied from over £16,000 in London to

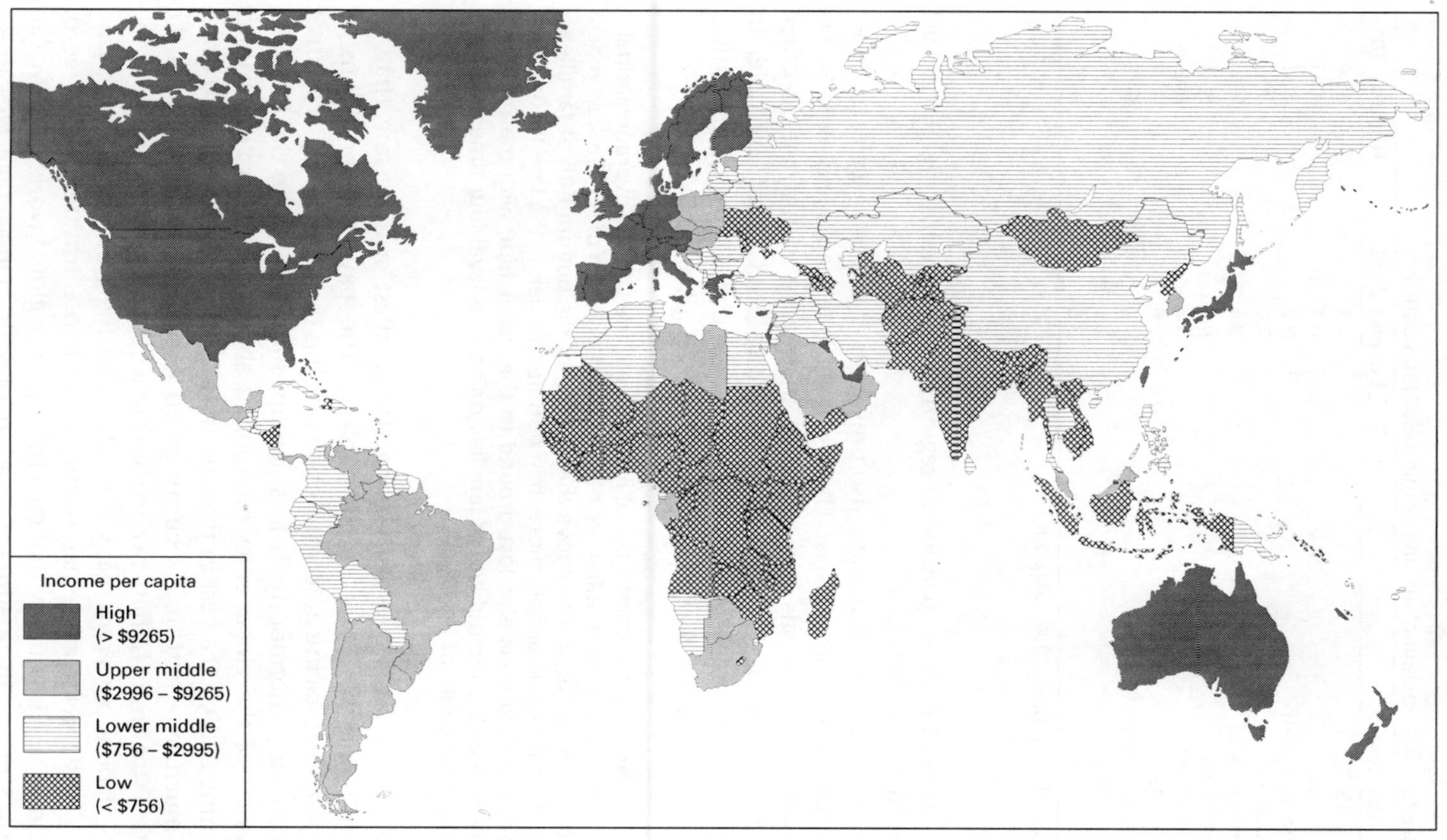

Figure 1.4 Distribution of low-, middle- and high-income countries

Source: Dicken, 2003: figure 17.1

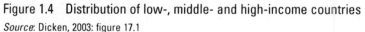

TABLE 1.3 Distribution of income within selected countries

Country (year)	Lowest 20%	Highest 20%
USA (1997)	5.2	46.4
UK (1995)	6.1	43.2
Germany (1994)	8.2	38.5
France (1995)	7.2	30.2
Sweden (1992)	9.6	34.5
India (1997)	8.1	46.1
China (1998)	5.9	46.6
Japan (1993)	10.6	35.7
Malaysia (1997)	4.4	54.3
Brazil (1998)	2.2	64.1
Mexico (1998)	3.5	57.4

Source: World Bank 2002, Table 2.8

less than £11,000 in peripheral regions such as Wales, the North East and Northern Ireland.

Income inequalities are also prevalent at the *local* scale, no more so than within large cities. The work of American sociologist Saskia Sassen (2001) reveals the extent of inequality within the so-called 'global cities' of London, New York and Tokyo. For example, in New York, income inequalities grew more quickly than in the United States as a whole. As Sassen (2001: 270) describes;

> New York has the worst income inequality in the US. The average annual income, adjusted for inflation, of the richest one-fifth of NY families rose to a level twenty times above the income of the bottom fifth of families in the 1996–1998 period, more than double this gap in the late 1970s. The city's middle class also lost ground to the richer fifth, with the median income of the top quintile of families rising to a level four times that of the middle fifth of families.

We can conclude by considering the smallest spatial scale that we identify in this book, the workplace. The pay and conditions of employment facing different workers within the same workplace can of course vary enormously. This is perhaps most extreme in the financial districts of the global cities mentioned above. An investment banker earning £200,000 a year in the City of London can be making 20 times as much as a part-time security guard or cleaner struggling to make £10,000. Although these two workers live in the same city, and both sell their labour power, in many ways they are living in different worlds.

This brief section has simply served to illustrate the degree of socio-spatial inequality that cuts through the global workforce. We can consider these inequalities at a variety of scales. On average, we can see

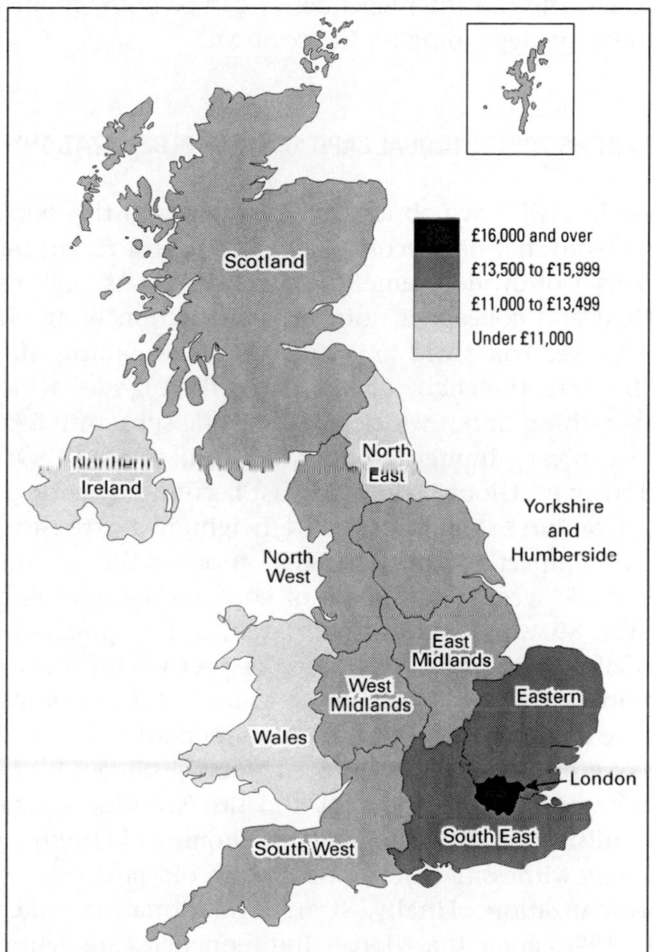

Figure 1.5 UK regional GDP per capita, 1999
Source: HM Treasury 2001: figure 1.1

that workers in North America are better off than those in sub-Saharan Africa, and that workers in Germany are more comfortable than their counterparts in Portugal. In the same way, workers in Southeast England tend to be better off than those earning a living in Northeast England. It is at the local scale, though, that inequities are lived and experienced. Even in the richest cities of the wealthiest countries, large numbers of workers are living in poverty. The uneven geography of wage work is fine-grained, and cuts deep. Critically, however, for our argument in this book, workers do not exist in isolation. Far from it; the fortunes of workers are in fact deeply interconnected at a variety of spatial scales, and this interdependency can be both empowering, and dis-empowering depending on the context.

STARTING POINTS: GLOBAL CAPITALISM OR GLOBALIZATION?

Astute readers will have observed that the title of this book does not contain a word that has become *de rigeur* in most recent publications on workers worldwide: namely, globalization. As political theorist David Held and colleagues note in their pathbreaking book *Global Transformations*, this word 'is in danger of becoming, if it has not already become, the cliché of our times: the big idea which encompasses everything ... but which delivers little substantive insight into the contemporary human condition' (Held et al., 1999: 1). We're inclined to agree. Globalization has fast become a chaotic concept. To be sure, it captures something of the heightened extensity, intensity, velocity and impact of human social relations in the new millennium. But the term's become 'a miasma of conflicting viewpoints' (Dicken et al., 2001: 89) and a battle-ground of rival interpretations. As Held et al. explain, there are at least three perspectives on the phenomena. 'Hyperglobalists', like management guru Kenichi Ohmae (1995), insist we're living in a qualitatively new era marked by the 'deep integration' of most places and peoples in globe-girdling sets of economic, financial and cultural relations. By contrast 'sceptics' – notably political economists Paul Hirst and Graeme Thompson (1996) – argue that we're merely witnessing a new phase of an old phenomenon: that is, internationalization. Finally, 'transformationalists' like Anthony Giddens (1990) argue that globalizing tendencies are definitely afoot, but that they have not as yet, and may never, produce/d a perfectly integrated global economy, polity and society.

We have no desire to adjudicate between these conflicting viewpoints. There is evidence for all three, which is why the debates on globalization have now reached something of an impasse. Clearly, something is happening 'out there' and, equally clearly, it is of relevance to workers worldwide. But in our view, using the term globalization as an analytical device will not get us very far in understanding the socio-geographic condition of labour at this moment in history. Instead we prefer the older term 'global capitalism' – one which had greater currency among social scientists during from the 1960s until the late 1980s. In our view this concept cuts more deeply. It tells us what we really need to know about the world in which we live and labour. That this is a predominantly capitalist world seems to us indisputable. We use the term capitalism in the classical Marxist sense to refer to a mode of producing goods and services predicated on the principle of 'accumulation for accumulation's sake'. There's scarcely a place on the planet where this mode of production does not have some purchase. Like Marx (1857: 43–4), we consider capitalism to be 'the general light tingeing all other colours and modifying them in its general quality'. With the late-1980s collapse of state communism in the former USSR and Eastern bloc, this system of production arguably now has few, if any, serious economic rivals.

This does not mean that capitalism is 'all there is'. Nor, in geographical terms, does it mean that capitalism is globalized in the sense that someone like Ohmae might use this term. One of the arguments of this book is that however 'global' some social relationships have become, place difference, uneven geographical development and local specificity persist. 'Global capitalism' is really a multitude of 'local capitalisms' that are connected by flows of people, goods and information. These flows are by no means uniform and they have different implications in different geographic contexts. So by the term 'global capitalism' we mean to register the fact capitalism is today the 'normal' economic system worldwide without implying that is globa*lized* in the sense of embroiling all places and countries *equally and uniformly*. Perhaps a better phrase to use, but for its inelegance, would be 'translocal capitalism', which signifies the distanciated but varied connections between a plethora of different places producing goods and services for profit. Some of these connections will be national, others international, and still others planetary in scale. So in cities like Milan or New York the latter will predominate, while in rural Nepal or the remote communities of Oman the connections are likely to be far less international and global.

While these are not exclusively economic connections, it is nonetheless precisely economic ties that in our view lie at the heart of things. The term globalization – which encompasses culture, politics, and even environment (see Held et al., 1990) – threatens to distract our attention from this important fact. Whatever real world changes the concept globalization describes, it remains the case that we live in a socio-geographic universe in which the rules of the game are distinctively capitalist. Interestingly, the massive public demonstrations against the effects of globalization – starting with the disruption of the World Trade Organization meeting in Seattle in 1999 – have become known as 'anti-capitalist' protests. Indeed, in the UK the recent May Day protests beginning in London in 2000 have been called 'carnivals against capitalism'. Within the capitalist universe the firm is the basic economic unit, the legal shell within which one or more workplaces produce, move, sell or maintain commodities or services. Together, firms comprise different economic sectors that, in turn, are the engine room of the capitalist economy. Our focus on capitalism does not implicate us in economic determinism: that is, the now discredited argument that the non-economic dimensions of our lives are subordinate 'in the final instance' to the economic one. Instead, we see capitalism as what Marx called a 'mode of life'. This means that the production of commodities in pursuit of profit intrudes into so many seemingly disparate elements of our existence and so many institutions that it's virtually impossible to understand the 'non-economic' domains in relative or absolute isolation. The capitalist and non-capitalist spheres mutually condition one another. We will see this numerous times on the pages that follow.

THE POLITICS OF A WORD: MYTHS OF GLOBALIZATION

All this said, we do not wish to dispense with the concept of globalization altogether. But our interest in it is as an *object of analysis* not an analytical device. David Harvey (1995: 1) has astutely asked us to consider 'why it is that the word "globalisation" has recently entered into our discourses ... Who put it there and why? How has the concept ... of globalisation been used ...?' These are good questions. They focus our attention less on how accurate the concept is as a descriptor of real world changes, and more on its use as a tool to make people *believe* we live in hyperintegrated world of ceaseless and irresistable flows. In the words of political scientist Barry Gills (1997: 12), 'globalization is an extremely political concept, even though it masquerades in a supposedly objective or purely analytical guise'.

This may seem an odd claim at first sight. After all, globalization is just a term used to describe a set of real world relations, events and processes. What, it might be asked, could be political about a mere word? And whose political interests might the word serve? The answers lie in the way that a specific set of meanings has become associated with the term globalization and how these meanings have been used by businesses, regulatory institutions and sometimes even workers themselves. As Dicken et al. (1997: 158) have observed, 'definitions are not mere semantic peccadilloes'. The meanings commonly evoked by the term globalization take the form of a set of myths. These myths have a grain of truth in them, but have been extended far beyond what is warranted by the factual evidence that brought them into existence in the first place. As such, they've seemingly taken on a life of their own, regardless of how accurate they are as depictions of contemporary socio-economic change. There are six of them we wish to consider and demystify. This book is, in an important sense, written *against* these myths.

The first myth is that we live in an increasingly 'borderless world' (Ohmae, 1990). This myth is particularly associated with those who wish to minimize restrictions on the movement of firms, reduce trade barriers and liberate financial flows. Globalization is often taken to mean that place matters less and less, what with the enhanced capacity of businesses, finance and people to either move around the globe or else 'collapse' space through virtually instantaneous communication technologies. As an ad by British Telecom once famously declared 'Geography is history'. However, as we've already intimated in this chapter, the notion of a 'placeless' world is a fiction not a fact. No matter how mobile some firms are, no matter how free to migrate some workers might be, no matter how much communication advances have shrunk the globe, place still matters for production, reproduction and consumption. In short, even globalization 'has to ground itself and be acted out in specific places' (Merrifield, 1993: 520). A second, and related myth is that globalization is some irresistible force that stands over and above different places and peoples. This myth has been a particular favourite among big business and several leading politicians. The refrain of Dr Richard Hu (1997), once Singapore's finance minister, is all too familiar: '... we have no choice but to open up and to compete in the world market to survive and prosper' (quoted in Dicken, 1998: xiii). More recently, Neil Holloway, chairman of Microsoft UK, has echoed these sentiments: 'Globalization', he declared, 'is going to happen and you can't stop it. You have to go with the momentum' (*The Guardian*, 6 April 2002). The imagery here is redolent of an incoming tide sweeping all before it: in short, one is told to go with the flow or else perish. It's an imagery that implies that places and people have little control over their fates, other than to adapt to globalization and hope for the best.

The third myth is that globalization signals the demise of the nation-state. No longer able to insulate their citizens and economies from outside forces, the argument is that national states can only modify, but not manage or control, transnational flows of people, information and goods. This is a myth because it is only partly correct, as we'll show in Chapter 4. While it is probably true that the *relative* powers of national states have altered in recent years, it is far too simplistic to talk of an absolute diminution of state power. Instead, there's arguably been a process of *re-scaling state power*, both upwards and downwards (Jessop, 1993). On the one side, national states have frequently devolved powers down to the local level, given how important sub-national differences now are in attracting and retaining outside investment. On the other side, though, the fact of heightened place-interdependence has produced the need for transnational state institutions to regulate spatial relations (a point we'll explore in Chapter 4). If we add all this together, it's probably more accurate to say that national states still do matter as regulatory bodies, but that they now function in a complex field of sub- and supra-national state power (Weiss, 1998).

This brings us to the three remaining myths, which concern workers specifically. The first of these is the myth of worker vulnerability. As Piven and Cloward (2000: 413) put it, 'Globalization seems to puncture a century old belief in worker power.' Bolstered by a dubious notion – the notion that firms are now hyper-mobile, while workers are hopelessly place-bound – this myth depicts globalization as a 'race to the bottom'. Firms are seen as being increasingly able to 'play-off' workers one against the other in a geographical divide-and-rule strategy. Relatedly, because trade barriers have been reduced in recent years under the auspices of the World Trade Organization (among others), there's a view afoot that workers worldwide are pitched into more open and intense relations of competition than ever before. Knowing that wage-workers in other places could produce goods more efficiently or cheaply, the idea is that globalization is forcing labour into so-called 'concessionary bargaining'. In this scenario, employers can compel workers in one place to agree a slimmed-down package of pay, conditions and workplace rights by using the threat of firm closure or relocation. For Peck (1996: 240) this is a bleak world of beggar-thy-neighbour behaviour – a world of 'hegemonic despotism' (this last concept is discussed critically in Chapter 5).

The fifth myth is directly related to the fourth: it's the myth of cheap labour. In this myth, firms with the capacity to choose between several possible production sites ultimately gravitate to places with the lowest labour costs. This myth goes back to the 1960s and 1970s, when a number of TNCs relocated factories to the Far East from Western Europe and North America marking the emergence of a so-called 'new international division of labour' (Fröbel et al., 1980). It's a particularly egregious myth because the locational decisions of capitalist firms are infinitely more complex than simply seeking out cheap workers. The skill levels, compliance, initiative and work-rate of labourers are all vitally important, as are the regulatory environment and relative location of the place being considered. Nonetheless, the myth of cheap labour exerts a powerful hold. It's become a shorthand for the 'war among workers' that supposedly prevails when the costs of labour vary so radically between places in the 'first', 'second' and 'third' worlds.

This brings us to the final myth, one that is found as much in the pro-worker literature as in the pro-business literature. In this myth, if workers are not to succumb to slash-and-burn globalization, then the suggestion is that they *must* 'up-scale' their actions to match the translocal forces that are undermining their well-being. In Turnbull's (2000: 383) words, 'Globalization often involves the image of ... workers ... being powerless in the fact of capital mobility unless they can themselves develop a global strategy in response.' The idea here is that if workers in different places join together they can act to prevent firms or certain regulatory authorities playing them off against each other.

To be sure, and as we'll see in Chapter 8, recent years have witnessed some innovative experiments in connecting different workers in different localities within and between nation states (see Wills, 1998). These have been aided by the existence of international and global trade union organizations, like the International Confederation of Free Trade Unions (ICFTU). However, these translocal experiments notwithstanding, the idea that wage-workers *must* today upscale their struggles for better pay, conditions and rights is a myth for the following reasons. To begin, it implies that locally-based worker actions are no longer necessary or sufficient. We argue otherwise, especially in Chapter 6. Additionally, this myth suggests that worker interests will normally be best served by joining with distant others elsewhere. Again, we argue that this is not necessarily so, since 'interests' are in fact scale-dependent rather than absolute. As Herod (2001: 6) rightly puts it, '"workers" interests may vary considerably depending upon the geographic context in which they find themselves'. Finally, the myth of up-scaling stakes workers' futures on what is, in fact, an often painstaking task of constructing inter-place alliances. Just because worker interdependence is arguably at an all time high, it does not mean that inter-place worker solidarity will *automatically* emerge as a result.

If we add our six globalization myths together we can see what a mutually reinforcing battery of beliefs they are. Though there's a grain of truth to each of them, the myths have over-reached their empirical veracity. Like all myths, they 'ignore the messiness of the world and the many inconvenient facts that this produces' (Jessop, 2001: 444). Consequently, their interest for us lies less in their in/accuracy, and more in the way they can serve particular social and geographical agendas. As noted above, Harvey (1995: 1) was among the first critics to ask 'how ... the conception of globalization [has] been used politically'. It's a very good question, for two reasons. First, it draws our attention to how concepts can have a certain life of their own, regardless of their factual accuracy. As Gibson-Graham (1996: 120) astutely notes, concepts can have a 'performativity' if promulgated widely enough, helping to create the realities they purport merely to describe. Second, Harvey's question allows us to consider just who stands to benefit if and when workers and communities buy-into the six myths we've discussed. An obvious response is that certain firms and pro-business regulatory institutions stand to benefit, because together the six myths put workers on the back-foot. They conjure up the image of globalization as a 'tsunami of transformation' (Herod et al., 1998: 3) in which workers have little power or voice, unless they make heroic efforts to organize themselves translocally. In this sense, the globalization myths are politically useful for those firms and organizations seeking to extract as much as possible from place-based labour. It's precisely for this reason that we draw critical attention to them here. For our argument in this book is that the prospects for wage-workers are not as dismal as the globalization myths would have us believe.

However, before we can make good on this argument we need to present an analytical framework to help us make sense of the opportunities and constraints for workers in our capitalist world. We begin, in the next chapter, by explaining the social parameters conditioning wage labourers, before, in Chapters 3 and 4, embedding these geographically. We should say at the outset that our theoretical framework is by no means comprehensive: no framework ever is. For instance, we do not say much about the recent work that questions 'the discursive construction of identities, ... workplace culture, symbols and representation of work and workers, and about sexuality and power in the production and reproduction of workplace inequalities' (McDowell, 1999: 146). We do not touch upon the work of those critical scholars who have begun to sketch out how workers use laughter, chatter and other tactics to resist the will of employers (for an overview see Auster, 1996). Nor do we consider how certain forms of work demand from individuals a certain form of emotional commitment – a smile, a laugh or a cry – or a certain aesthetical commitment – a particular dress, haircut or 'look' – all of which means workers are giving ever more of themselves in the name of work (Auster, 1996). However, Part One should equip readers with the basic tools to comprehend what happens to wage workers in the modern world, how they respond and why geography matters in the process.

Part One

GEOGRAPHIES OF LABOUR

two

The Social Relations of Labour: Working in a Capitalist World

Wage labour is a profoundly *social* activity. It is anything but a dry, technical process of assembling materials, making goods or delivering services. Even the most mundane, repetitive act of production implicates wage labourers in a number of 'vertical' and 'horizontal' social relationships with employers and fellow labourers respectively. It also entails links with groups and institutions outside the workplace, not to mention a range of non-capitalist social differences among people relating to gender, ethnicity, age and the like. It is this constellation of social relations, institutions and identities that this chapter explores. We start by identifying a set of commonalities among wage-workers before step-by-step exploring lines of social difference and division. In reality, the social dimensions of wage-work cannot be considered separately from the geographical dimensions. However, for the sake of clarity we defer consideration of the latter until Chapters 3 and 4.

DEFINING CAPITALIST LABOUR: WORKERS AS PSEUDO-COMMODITIES

If we abstract from the evident differences between millions of workplaces and firms worldwide we start to see the underlying logic that governs their activities. This logic is what we routinely call 'the capitalist system'. This system doesn't exist outside the myriad industries, workers and state bodies that together constitute it but is, rather, the precondition and result of their activities. It takes the following form:

$$M \ldots C \begin{cases} MP \\ \\ LP \end{cases} P \ldots C^{*} \ldots M + \Delta$$

In this system money (M) is advanced by owners of firms (capitalists) to purchase commodities (C) of two kinds, namely MP (means of production: buildings, machinery, raw materials etc.) and LP (the labour power of workers). This results in a process of production (P) leading to a new commodity or commodities (C*) which are then distributed and sold to consumers for the original sum of money advanced plus a profit (Δ). Part of this profit is then reinvested into further rounds of production and so on in a ceaseless process of accumulation.

There are several key things that distinguish this mode of producing goods and services from others past and present. First, capitalism is fundamentally growth orientated. Its goal is to generate more and more profits; all other ends are subordinate to this overriding one. Second, capitalism is an inherently competitive economic system. Within different economic sectors numerous firms constantly jostle to sell their commodities to consumers. Only in oligopolistic sectors can firms moderate the chill winds of competition. Third, this competitive ethic sets up powerful incentives for businesses to innovate their products and production processes. If these can be made cheaper or better then firms stand to make more profits and beat out their economic rivals.

So *growth*, *competition* and *innovation* are the life-blood of capitalism. It's a profoundly dynamic, restless way of producing things. What has this got to do with labour? As the diagram above shows, labour is one input into the production process. It's one means to the end of capital accumulation. More specifically, it takes the form of a *commodity*. This is what distinguishes labourers within a capitalist system from those in non-capitalist systems. All commodities have a use-value (that is, a practical function) and an exchange-value (that is, a price). The use-value of labourers is their capacity to work – to perform workplace tasks at certain skill levels for certain periods of time – in return for which their employers remunerate them. Unlike, say slavery, a capitalist does not buy rights to the person but to the person's capacity to work.

This brings us to the difference between capitalist wage labour and 'true' commodities. Most commodities are non-sentient. Be they cars, computers or paper clips, they are typically material artefacts that can be made and disposed of as companies see fit. But labour is different: it is really a *pseudo-commodity*. As Michael Storper and Richard Walker (1989: 155) put it, 'Labour differs fundamentally from real commodities because it is embodied in living, conscious human beings …'. This has three implications. First, it means that labourers are only *temporarily* commodities – except in those extreme and disturbing cases of bonded or indentured wage-workers (Bales, 1999). Each working day they assume the *form* of a commodity, but this does not alter the fact that they were not born and raised to *be* commodities. Wage-workers, because they are people and not tables or chairs, require happiness, sustenance, shelter, entertainment, good health and all the other ingredients of a life. This means that, unlike most other inputs into capitalist production, they have irreducible needs and a plethora of wants (though these may not always be met in practice). Second, because wage labourers are physiologically and mentally complex beings capable of independent thought and action – again, in contrast to most conventional commodities – they have *agency*. In the words of Ray Hudson (2001: 220), 'Workers are active subjects, not passive objects …'. That is, within certain constraints, they can determine how well, for how long and under what conditions they are willing to work for employers (the sociologist Anthony Giddens has famously theorized social agency; we discuss his idea in Chapter 6). Finally, this means that wage workers – yet again, unlike other commodities – necessarily enter into a *social relationship* with their employers. In Marxian parlance, this is a *class* relationship, since workers' pseudo-commodity status is what distinguishes them as a social group from the relative minority of capitalists who purchase their labour power. It is a relationship that exists in the workplace and outside it, the combination of which defines the *labour market*: that is, the quantitative and qualitative matching of labour supply with the demand for labour among employers. This matching can occur directly between workers and employers or via so-called 'labour market intermediaries' who act to match workers with employers.

SITUATING LABOUR IN THE CAPITALIST PRODUCTION PROCESS

Understanding labour's distinctiveness as a commodity helps us to understand its special place in the capitalist production process. Labour is the only commodity input to production that employers cannot treat as an object. The employee–employer relation that develops because labour is *embodied* is simultaneously necessary and problematic, at once a requirement and an obstacle. Another way of saying this is to observe that the relationship between employers and actual or potential employees in capitalist societies is a *co-dependent one of cooperation and conflict*. The two are indissociable; they are sides of a coin. But why is this?

The reasons have everything to do with how the 'rules' of capitalism position employers and employees differently in terms of their aims and objectives. Capitalists aim to make money: this is their *raison d'être*. From their perspective, workers are both a *cost of production* and a necessary *input to production*. Even today, wages are the highest single cost for most businesses worldwide. Meanwhile, it is neither possible nor desirable to eliminate wage-workers from most production processes. Some production tasks – such as designing aircraft or building houses – are either unamenable or only partly amenable to being done by machines, computers and the like. As Storper and Walker (1989: 160) argue, 'Labor, as the [fundamental] factor in production ..., can never be entirely replaced'. In any case, at the aggregrate level, if all capitalists were to significantly reduce their labour forces then they would lose their biggest market: namely, ordinary consumers (since workers not only earn money but spend it too). For these reasons businesses of all stripes require a sufficient number and calibre of workers both to undertake certain production tasks and to purchase commodities as part of their (and their families') reproduction.

Yet all this said, for individual firms operating in a competitive environment, it is 'rational' to pursue some or all of a range of strategies that are not necessarily in the interests of all employees. Some of these will be workplace strategies (see Box 2.1). One is cutting or holding-down wages; another is displacing existing workers with more cost-effective machinery or a more effective division of tasks within the firm; still another is to exercise maximum control and surveillance over the workforce so that it performs at the required standard for as much of each working day as is possible. Some strategies are associated with adjusting the 'boundaries' of particular workplaces, either through the shifting of tasks to other workplaces within a firm, or through the outsourcing or subcontracting of tasks to workers in other firms. There are a variety of other strategies available, many of them pursued outside the workplace (such as groups of firms pressuring governments to reform labour laws in their interests). Together they comprise what Piven and Cloward (2000: 149) call the 'power repertoires' of capitalists. These vary between firms and economic sectors, as employers seek to get the most out of workers for the least cost (which is not necessarily to say cheaply). They come into play both when business is good (as in the Dyson case) and when it is not so good (as when the telecom giant Motorola suddenly laid off some 3,000 of its Scottish workers in mid-2001 because of oversupply in the European mobile phone market (*The Guardian*, 25 April 2001)). And they are undertaken by firms individually (as in the Dyson case) or collectively (for example through employer organizations like the UK Confederation of British Industry).

Box 2.1 Employment relations in the workplace

Employment relations in the workplace – and outside it – are two-way or 'bilateral'. In all workplaces, there are employer–worker tensions between:

Worker autonomy	Management control
Training and skills development	No training, low skill
Good wages	Low wages
Fixed and few work hours	Long and variable hours
Promotion opportunities	No promotion prospects
Job security	Job insecurity
Full-time status	Part-time or temporary status
Non-wage perks (e.g. a company pension)	No non-wage perks
Good workplace facilities	Poor facilities

Employers tend to emphasize the right-hand attributes when they seek to reduce labour costs and those on the left when they wish to increase labour productivity. Additionally, within a single workplace, or among workplaces within a single firm, some workers may be subject to different treatment at the same time depending on their place within the production process. (Adapted from Gough, 2001: 19).

The precise balance of sticks and carrots that firms use to secure the right workers at a suitable price performing at appropriate productivity levels is a contingent matter. It always depends in part on how far workers are willing and able to comply with the demands of employers. On the one hand, workers *need* employers. In the contemporary world, the opportunities for non-waged employment are relatively few compared with those for paid work. Though we rarely stop to think about it, most people simply have little or no choice beyond their childhood and teenage years but to offer themselves as wage labourers. Indeed, a good deal of most people's early life is spent doing things – like going to school – that help prepare them for a life of wage-work. The young are, in effect, 'wage-workers-in-waiting'. But this does not mean that potential and actual workers lack choices or are completely at the beck-and-call of capitalist companies (though, as we'll see later in the book many of them are in certain times and places). As we've already observed, labourers have agency. The fact that employers need workers (as much as the other way around) gives the latter a degree of potential control over the terms and conditions of their labour. Workers too have 'power repertoires'. These can be utilized inside or outside of the workplace to either pressurize (for example, by going on strike) or support (for example, by being willing to be retrained or relocated) their employers (see Box 2.2). And, mirroring the repertoires of the latter, they can be pursued within or between individual workplaces and firms. Trades unions have, historically, been the major institution through which these worker repertoires have been channelled (Box 2.3).

Box 2.2 Power repertoires

According to Piven and Cloward (2000: 413), 'power repertoires' are 'historically and geographically distinct power strategies employed by capital and labour separately and together'. Putting spatio-temporal specificity aside for one moment, we can distinguish different types of power strategies pursued by employees and their wage-workers. First, we can distinguish *physical* from *non-physical* strategies. Examples of the former are new machines that are more productive than workers or strikes against employers. In the latter case, persuasion is involved by one party to gain consent from the other over some new arrangement (for example, a wage freeze or shorter working hours). Second, we can distinguish *workplace* from *non-workplace* strategies. Examples of the latter include companies and labour unions pressuring governments for changes in employment law. Third, we can distinguish *incremental* from *non-incremental* strategies. In contrast to the former, the latter are strategies that are 'sprung' suddenly, such as the earlier mentioned decision by Motorola to lay-off 3,000 Scottish workers. They give the other party little time to think or act. Fourth, we can distinguish *official* from *unofficial* strategies. The first of these are well-established and legal strategies, the second more original and/or of dubious legality. Workers sabotaging machinery or employers breaching minimum wage legislation are examples of the latter. Fifth, we can distinguish *reactive* from *proactive* strategies. The former are resistance strategies deployed when the other party has the upper hand, the former control strategies when the other party is relatively weak. Finally, and for workers only, we can distinguish *accommodatory* and *transformative* strategies. The former are strategies that aim to get workers the best possible deal within the limits of a capitalist world economy. The latter, more radically, attempt to take workers out of capitalism altogether. In theory, this can be done via a revolution (see the comments on this in the Preface) or, less ambitiously, by groups of workers 'de-linking' from capitalism and seeking non-waged ways of making a living while capitalism lives on (see Chapter 6). In all cases, the power of employers and employees are able to exert over one another is a relative not absolute affair. As Held et al. (1999: 20) argue, 'Power has to be understood as a relational phenomena.' Thus a strategy that worked for employers or workers in one place and time may not achieve the same results elsewhere or later. See the introduction to Sharp et al. (2000) for a survey of theories of power and resistance.

Box 2.3 Employer and worker organizations

In order to formalize the employer–employee relationship within and between workplaces, firms, and industries at a variety of scales, both sides of the relationship have developed their own organizations and institutions. On the employer side, there are two key types of institutions. First, there are those that bring together the common interests of a broad range of businesses in a particular place. These may be locally oriented, such as Chambers of Commerce, or nationally focused as with the UK's CBI (Confederation of British Industry). Second, there is a broad range of sector-based trade associations such as the UK's Computing Services and Software Association (CSSA). While typically nationally based, such organizations may also be members of international organizations. For example, the World Information Technology and Services Alliance (WITSA) is a consortium of over 40 IT industry associations from around the world. On the worker side, three types of organizational structure are prominent. First, individual workers may be members of professional institutions, associations or guilds that represent and accredit their particular talents or skills. An example is the Royal Institute of British Architects. Second, within the workplace, various forms of worker's committees provide an opportunity for employees to present a collective viewpoint to management. Formal 'Works Councils' have become increasingly prevalent and may be networked together to enable information sharing in multi-plant firms, as in the case of the European Works Councils described by Wills (2000). Third, and most importantly, trade unions of different forms provide a crucial vehicle for collective organization.

It should be noted that neither capitalists nor workers are free to pursue all the power repertoires that are theoretically at their disposal. First, there are *structural constraints* in play. These are barriers to action posed by prevailing macro-economic and institutional arrangements. For example, a state may make it illegal for workers to strike unless certain stringent conditions are fulfilled. Second, it is important to note the importance of what critical theorist Antonio Gramsci famously called *hegemony*. Hegemony describes how powerful social groups seek to pursue their agendas by winning consent from subordinate groups. Hegemonic ideas seem acceptable because people have been socialized to think about them that way, not because they are inherently decent or fair. For instance, the idea that workers must submit to the forces of globalization – dealt with in the Preface – is a hegemonic idea because it has come to seem normal and unexceptional. In this way, hegemonic ideas help circumscribe 'permissible' and 'achievable' actions by employers and employees. Put differently, hegemony works through *discourses* that become taken-for-granted in society. Thus, for example, if globalization seems irresistible to certain workers, then rather than take a proactive stance in the face of employer pressure to freeze wages, they may simply capitulate.

Again, the balance of pressure and support varies from case to case. But underlying this variable geometry of cooperation and conflict, workers do share some common interests by virtue of where they are positioned within capitalism. Three loom large:

- **Firstly** they need *at a minimum* to earn a 'living wage'. This is the remuneration required to enable basic physical and social reproduction of workers and their dependents. The living wage varies over time and space because the cost of living does. As we'll see in Chapter 6 living wage campaigns have occurred in a number of places, uniting workers with other local groups.

  The question of workers' dependents emphasizes the profound importance of seeking and securing paid employment in a capitalist world. Without a salary neither workers nor their husbands, wives, lovers, children or other relatives have the means to live anything other than a meagre existence that may, in turn, undermine their future employment prospects. Neither, for that matter, would the mass of retired workers with pensions have much of a living. The reason is that pensions are partly funded from *current* wages. And yet one of the 'logics' of capitalism is to generate short- and long-term unemployment! As already noted, either firing existing workers or hiring fewer future workers is often a rational strategy for businesses to pursue in order to survive and/or prosper in the short-to-medium term. So what we're saying is that in capitalist societies there's a common thread uniting four social groups, only one of which works: namely, paid labourers, their dependents, retirees and the unemployed. To a greater or lesser degree, all are *directly* or *indirectly* dependent upon earning a living through the sale of labour power. If we revisit our earlier diagram it now looks something like this:

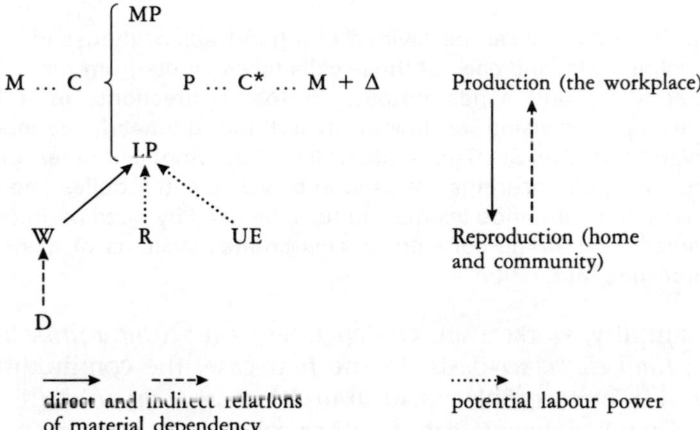

where W is workers, D is non-working dependents, R is the retired and UE is the involuntary unemployed. These work/non-work links simply reinforce the point that paid labour is the glue that holds capitalist societies together.

- **Secondly** over and above this basic dependence upon earning a salary, it is in workers' interest to maximize all the potential benefits they can derive from their employment position. This is not just a question of improved wages, but also of rights (such as the right not to have to work long hours) and related non-monetary entitlements (like job training, company health care, etc.). The average employer is considerably wealthier than the average worker: after all, they appropriate the lion's share of firm profits. When the 'wealth gap' between workers and employers becomes too stark, some form of protest among the former typically follows. Typically, the higher up the employment ladder workers are able to climb the more likely they are to extract a generous bundle of monetary rewards, work rights, and job entitlements from their employer. Why? Because workers who have high level or hard-to-find skills are, by definition, scarce pseudo-commodities. They can thus command superior pay and conditions, at least until and unless they are surplus to a firm's requirements.
- **Finally** across the social division of labour mentioned earlier, workers are directly and indirectly dependent upon one another for their jobs. This is a second ('horizontal') bilateral relationship to add to the 'vertical' one workers have with their employers. To understand this inter-worker dependency, we can think of the capitalist 'system' as comprised of millions of network relations between different units of production (Yeung, 1994). As Figure 1.3

illustrates, only the simplest of economic activities can avoid entraining numerous firms employing numerous workers. To quote Sayer and Walker (1992: 111):

> ... the economy can be divided into hundreds of thousands of production cells, and each of those cells takes inputs from many [workplaces] ... and sends outputs in many directions; in a highly developed division of labour everything ultimately connects to everything else. Steel goes into drill bits, automobile frames and ball-bearings; ball-bearings are used in bicycles, automobiles and textile machinery; automobiles may, in turn, be used by pizza parlours, gardeners and airline companies. This creates systems of nested and branched integration.

To simplify, workers are co-dependent on an *intra-firm, inter-firm* and *final consumer* basis. In the first case, the commodities produced in one workplace within multi-plant companies (like TNCs) usually travel downstream to other parts of the firm. In the second case, inter-firm ties – such as subcontracting arrangements – involve similar highly integrated commodity flows. In the third case, and more indirectly, when workers act as everyday consumers they purchase commodities made by fellow workers they'll likely never know or meet. In each of three cases, then, workers in one part of a network depend on others – acting as labour power or consumers – to process, assemble or purchase goods or services leaving their workplace.

Of course, when one adds together what, in an ideal world, employers and employees want of each other one is faced with a zero-sum game. En masse, workers could never, say, labour for a day or two per week for high wages, just as firms could never, say, force their workers to labour hard 20 hours each day for little reward (unless they were wage-slaves: see Bales, 1999). As Storper and Walker (1989: 168) rightly argue, 'because employers and workers are captives of each other ... neither side is free to get all it wants ... [P]roduction is [thus] successfully undertaken within an unfolding process of negotiation.'

CAPITALIST–WORKER RELATIONS: A FIELD OF TENSIONS

The capitalist-worker nexus is clearly a tense and contradictory one. As we've seen, it spills out beyond the workplace to the sites of worker reproduction such as the home. Having identified the general class interests of employers and wage-workers (along with ancillary groups like the latter's dependents, retirees and the unemployed), we can now specify precisely what the to-and-fro of consent and conflict occurs over. From the perspective of capitalist firms, a 'perfect' labour market would exist when just the right number of workers of the right characteristics were available for the right wages working under conditions dictated by the firm. However, as Peck (1996: 26–40) argues, in reality four problems intrude into this idealized equation of labour supply, labour demand and workplace control:

The problem of labour incorporation

Men and women do not have children in order to fill slots in the labour market (or at least not usually). Capitalist firms thus do not have much control over the type and volume of labour supply or the willingness of potential workers to become actual workers. This is left to the vagaries of things irreducible to capitalist class relations, such as family values, state policies on childbirth, the size of the population, the nature of the education system (if one exists), and so on. If wage-work was the exception rather than the norm; if too few people were born; if only the over 40s were allowed to undertake paid work: if these and other constraints on labour supply applied then capitalist firms would confront major problems in incorporating a sufficient number and type of workers into the capitalist system. Likewise, potential workers would have a hard time surviving unless non-capitalist ways of making a living were readily available.

The problem of labour allocation

As the five stories told at the start of Chapter 1 illustrate graphically, different types of labour need to be slotted into different firms and sectors of the capitalist economy. These variations of labour relate not just to the jobs performed or the skill levels required but also the *social characteristics* of the workers involved. From an employer's perspective, men rather than women, blacks rather than whites, gays rather than heterosexuals might be better doing some jobs. Thus, for capitalists, in a perfect labour market the right *types* of workers would be matched to the right job *categories* – no more and no less than were required. However, as will be explained later, because firms have limited control over how workers are assigned social labels outside the labour market, allocation can be a problem. Even if it were not, in practice some groups of workers might not like being offered only certain restricted kinds of employment opportunities. Additionally, because firms may over time alter their labour demands in response to competition, then the existing labour force available may not be of the right quality or quantity. Of course, this latter scenario is a problem for workers, who may face unemployment or the prospect of forced migration as a result.

The problem of labour control

The two problems discussed above are supply and demand problems. The third is a workplace issue. Workers are hired because of their perceived performance capacity, which is their potential to do certain jobs at a certain level for a certain wage for a given period of time. Whether this potential is realized to an employer's satisfaction depends on a number of things: the complexity of the job; the degree of control exerted by employers in the workplace; the work ethic labourers have instilled from their schooling, family and friends; the legal rules governing how many hours labourers can work and under what conditions; the power of workers' organizations (like trades unions) and so on. Though employers might, in an ideal world, want total control of the labour process they cannot achieve this. In part this is because workers often have the agency to demand some say in how long they work, for how much, and in what physical surroundings – especially at the upper levels of the occupational ladder. But it's also because many of the ingredients that go in to making potentially or actually 'compliant' and 'productive' workers are to be found outside the workplace.

The problem of labour reproduction

The final problem is a short- and long-term labour supply problem. Employers need workers to be sufficiently happy and healthy for them to work well. They also need them to possess the required skills, aptitudes and attitudes to undertake their assigned jobs. All these things are influenced by what goes on in the reproductive spheres of home and community. As with the three problems already mentioned, firms have only partial control over reproduction. Once workers leave the workplace they are 'free'. It takes time, money and the non-paid labour of others (like family members) to reproduce workers day in and day out. Employers cannot on their own or together coordinate reproduction activities so as to get the desired outcome: that is, productive and appropriately qualified workers. Conversely, though workers do not use their reproduction time simply in order to prepare themselves for work, they need the requisite bundle of reproduction activities and resources to enable them to gain and hold down a job. Yet even though they have more control over reproduction than employers, they are not able to perfectly dictate its rhythms. In any case, some reproduction activities simply contradict the demands of wage labour. Think, for example, of the worker who gets divorced but retains custody of his/her four young children. That worker must labour to support the children, yet may find the daily non-work time spent raising them so exhausting that his or her absenteeism from work increases. Finally, current wage-workers aside, there's the necessity for the retired and the unemployed to reproduce themselves on a short- and long-term basis. If retirement incomes are inadequate or irregular, and if the unemployed suffer the combined effects of no income and few job prospects, then this can threaten the entire social fabric.

What conclusions can we draw from all this? The principal one is this: neither employers nor workers (nor the ancillary groups we've mentioned) can satisfy their individual and group interests within the confines of capitalism alone. Left to their own devices, the 'rules' of capitalism and the class relations intrinsic to it do *not* provide a sufficient means for the two major parties involved (and those indirectly involved) to realize their economic and other ambitions. In the absence of non-capitalist institutions and processes these rules and relations would, in fact, produce what labour sociologist Antonella Picchio (1992: 108) has called 'continual and often lacerating conflicts ...'. Quite simply, employers and workers would make demands on each other that neither, on their own, can deliver. At the same time, capitalism's inherent tendency to generate short- and long-term unemployment would, left unchecked, produce a pool of deeply unhappy individuals and families. In other words, there simply cannot be a 'pure' market in selling, buying, using and firing labour power. The supply of, and demand for, wage labour, along with what goes on in the workplace, require two additional things. The first is a raft of *non-capitalist institutions* that can help to regulate labour incorporation, allocation, control and reproduction. The principal institution here is the state. The second is a raft of *intra- and non-capitalist social relations and identities* that help to differentiate and unify workers. These two things comprise – and complicate – the wider social context in which wage-workers sell their labour power (Figure 2.1). They serve to place limits on the dynamics of capitalism and class relations, while cleaving the apparently coherent interests of

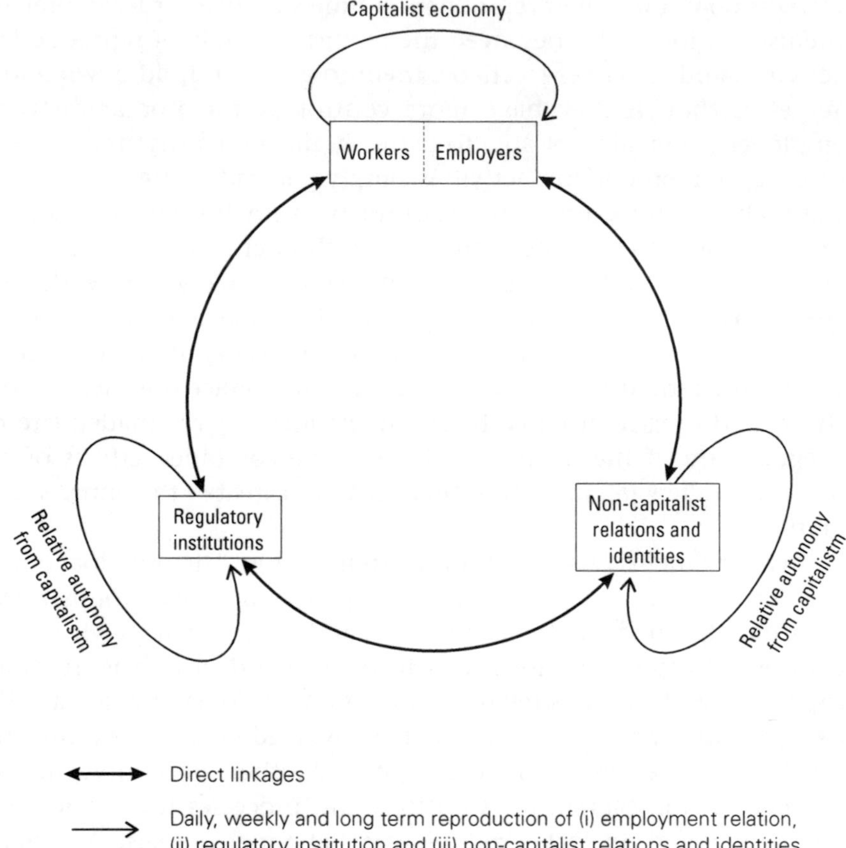

Figure 2.1 The social context for the sale and purchase of labour power

employers, workers, dependents, retirees and the unemployed. Together, as we will see in the remaining sections of the chapter, they can serve to enhance or hinder the power repertoires that employers and workers can deploy in their interactions within and beyond the workplace.

THE SOCIAL REGULATION OF THE EMPLOYER–WORKER NEXUS

If wage-workers and employers are to contain the contradictions of their relationship within tolerable limits, if those reliant on others' wage-labour are to reproduce themselves, and if unemployment is not to periodically reach crisis proportions, then *social regulation* is

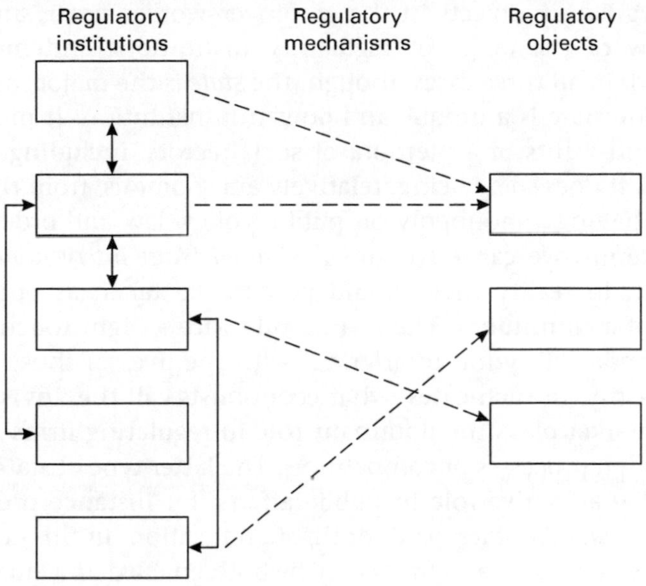

Figure 2.2 Aspects of social regulation

required. The word 'regulate' means to put a thing or a relationship on a consistent footing. In the case of labour, regulation is 'social' because it deals with real people enmeshed in multiple social relations that are conditioned by sets of societal institutions. As we'll now explain, both capitalist firms and their workers are socially regulated, along with those groups outside the labour force. This regulation is both tacit and explicit; it is sometimes unintentional, sometimes planned. Regulation, if you like, provides a second set of 'rules' and expectations governing what is possible and not possible for employers and employees. These rules and expectations 'mesh' with the ones intrinsic to capitalism. But they do not always do so perfectly or intentionally. The social institutions that provide regulation may not always be able to curb the potential problems that capitalist labour markets throw up. Many of them were not created to be regulatory institutions but have taken on a regulatory role over time. Indeed, these institutions may even create unintended problems of their own.

All social regulation involves *objects of regulation*, *institutions for regulation* and *mechanisms of regulation*. In theory, many different institutions can be created that use multiple mechanisms to regulate the same or different objects (Figure 2.2). Below, we identify the three main regulatory objects in the employer–worker nexus and give an overview of the range of regulatory institutions and mechanisms involved. In all three cases, though, the *state* is the major institutional actor. The state is a unique and powerful institution. It mediates the needs and wants of a plethora of social actors, including firms and workers. It does so by being 'relatively autonomous' from these actors and by having a monopoly on public policy, law and order. In theoretical terms, we can distinguish *minimal* from *interventionist* states, although in reality they should perhaps be taken as opposite end points of a continuum. The former rule with a 'light touch' and take little responsibility for 'interfering' with the lives of those subject to its authority. In such cases, what economists call the 'invisible hand' of the market plays the dominant role in regulating firms and workers through processes of competition. The latter type of states, by contrast, play an active role in public affairs, for instance providing an elaborate 'welfare state' paid for through taxation. In this context, the activities of workers and firms may be both enabled and circumscribed by state regulations in a wide variety of ways. Both types of state are compatible with either *democratic* or *dictatorial* modes of governing. The former involves the public electing those who will control state institutions for a finite period; the latter does not entail such democratic choice (the former USSR is a good example of a dictatorial state). Finally, it is important to know that *quasi-state* bodies can exert almost as much power over everyday affairs as state bodies do. Quasi-state bodies are not formally part of the state but are authorized to act on a state's behalf. A good example, at the global scale, is the World Trade Organisation (WTO). This body applies global trade rules on behalf of the national governments that comprise its membership.

This last fact indicates that in reality, we need to talk about states in the plural, since there exist local, national and supranational state and quasi-state bodies, each with their own powers and responsibilities (for instance, Manchester, England where three of us work, is governed by local politicians, who are subordinate to the British government in London which, in turn, is subject to European Union law). We also need to recognize that the precise composition of state apparatuses is historically and geographically variable. For now, though, we'll talk about the state in the singular in order to tease out theoretically the central regulatory foci and responses that preoccupy capitalist states in practice. We also employ an ecumenical definition of the state that encompasses its legislative (policy making), administrative (policy implementing), productive (goods and services producing) and legal (rule enforcing) arms. When, in the rest of this book, we use the word 'institutions', we are generally referring to those discussed in this section.

Regulating capitalist firms

The state is the most powerful regulator of economic activity in capitalist societies. According to Marxist and neo-Marxist state theorists like Hudson (2001: Chapter 3), it both internalizes and seeks to manage the contradictions of capital accumulation. To perform this managerial role the state must be *formally* independent from the economy. It cannot be seen to 'side' with any one economic group at the expense of another. Yet, at the same time, the state cannot be *substantively* independent, a 'political' domain somehow detached from capital accumulation and class relations. In effect, the state is a 'relatively autonomous ... "mediating" agency' (Hudson, 2000: 50). It seeks to balance the needs and wants of capitalist firms and the needs and wants of wage-workers, their dependents, the retired, and the unemployed. In so doing, its two principal functions are, first, to ensure the material conditions are right for continued economic growth (an *accumulation* function) and, second, to prevent widespread discontent about the capitalist system (a *legitimation* function) (Habermas, 1976). In this sub-section we consider the state–business relationship. In the next two we look at state (and non-state) regulation of those depending directly or indirectly on the sale of labour power.

Hudson (2001), summarizing over 20 years of Marxian theorizing about the capitalist state, argues that it takes on two related roles vis-à-vis firms and, by implication, the economy they collectively constitute. First, the state sets what he calls *framework conditions*. These are sets of formal rules and regulations that place limits on what firms can and cannot do – both in relation to each other and to their workforces. If you like, they are the political parameters within which employers must operate. Company laws, labour laws and pension rules are especially important here, and are devised through negotiation with workers, employers and others. These laws uphold company, worker, non-worker and retiree rights, while preventing the worst excesses of firm behaviour – such as, for example, employing under-age workers or undermining competition through monopolistic practices. To be effective, of course, framework conditions must be enforceable, and weak states that are low on resources and authority may be unable to achieve this.

First, states can intervene in firm behaviour more overtly. As Hudson argues, this has two aspects. The first is *direct provision of economically necessary goods and services*. Collectively, firms require things that they are unable to provide individually. Good examples are extended transport networks, power systems, water supply networks, liveable environments and education systems. Though it's not impossible for groups of businesses to produce these major goods and services for society as a whole, the state has the resources – through general taxation, for example – and possesses the overall responsibility to provide at least some of these 'public services'. 'Public sector' bodies like Britain's National Health Service and the United State's Environmental Protection Agency usually run these services. Second, states also seek *to directly influence the volume and type of firm activity within the economy*. They do this at the macro-economic level by controlling things like money supply, interest and exchange rates, customs duties, and taxation. But they also do it at the micro-economic level by, for example, building purpose-made facilities for particular firms, offering certain industries tax breaks, or establishing a free advisory service for business start-ups. Again, specific state institutions may take responsibility for delivering these various regulatory mechanisms.

Quasi-state bodies can be involved in all of these state activities. Many of these bodies are not directly elected by the general public and yet can be very powerful. These are institutions set up by the state and given varying degrees of autonomy in their actions, which may operate at the subnational, national or international scales. There are several interrelated reasons as to why quasi-state bodies may be established. First, their formation may simply be a means for states to act in ways they could not otherwise. For example, the WTO allows states to act together to regulate international trade. Second, their establishment may reflect attempts to give certain non-elected groups within society – such as businesses or civil society institutions – a greater role in decision-making and regulatory practices. In the UK, for example, over the last 20 years there has been increased private sector involvement in areas such as urban regeneration and workforce training. Third, they may be established to circumvent or challenge existing regulatory institutions or practices that are not achieving the required results. For instance, an urban regeneration body may be given powers that supersede those previously held by local government. Fourth, quasi-state bodies may be created to try and 'de-politicize' certain forms of regulatory activity where direct state intervention is deemed inappropriate or undesirable. The regulation of certain sectors dominated by a few large firms may be devolved to a quasi-state body, for example, so that the state is not seen to be allying with individual companies when taking regulatory decisions.

Regulating wage workers

Just as the state is the major regulatory institution for business so too is it the principal regulator of potential, actual and former wage-workers. This has four dimensions. In the first place, the state is a *major employer*, especially in Western countries that possess the most elaborate state apparatuses. A staggering 300 million workers are currently employed in state institutions worldwide. In Britain alone the state spent some £96 billion in 2000 on its employees (GMB, 2002). Of course, these workers are not guaranteed special rights and rewards simply because they are public rather than private sector workers. But they are often shielded from the rigours of the capitalist labour market, especially if they are employed in essential state institutions like the police force, the education system or the fire service. Second, in the private sector, the setting of framework conditions and the two forms of direct intervention identified by Hudson clearly affect the level, type and quality of employment economy wide. An important element of this for workers is the way the state guarantees (or fails to guarantee) basic employment rights and entitlements in both the public and private sectors. Third, outside the workplace, as Gough (2001: 27) observes, the state takes responsibility for *particular aspects of worker reproduction* that would be poorly achieved without state intervention. Just as capitalist firms together require infrastructural goods and services they would be hard pushed to deliver without state assistance, so too do workers have reproduction requirements that the state is uniquely placed to satisfy. Three of these loom large: health care, housing and job training. None are cheap for workers to acquire. In each case, economies of scale are achieved when these are provided to many workers by the state simultaneously. Finally, the state is the major regulator of *migration flows*, especially across international borders. National states can directly control the number and timing of outside immigrants, unless circumstances or a lack of resources dictate otherwise. Again, quasi-state bodies can have an important regulatory role to play in each of the four cases.

So much, then, for the state. Though a key regulatory institution, it is by no means the only body that regulates wage-workers (and related social groups) in the productive and reproductive spheres. If we cast an eye over what political theorists call 'civil society', we find all manner of regulatory institutions that are neither businesses in the normal sense nor part of the state apparatus (Figure 2.3). On the one side, there's the family. Of course, many workers are single, but at some point in their lives virtually all workers are socialized within a

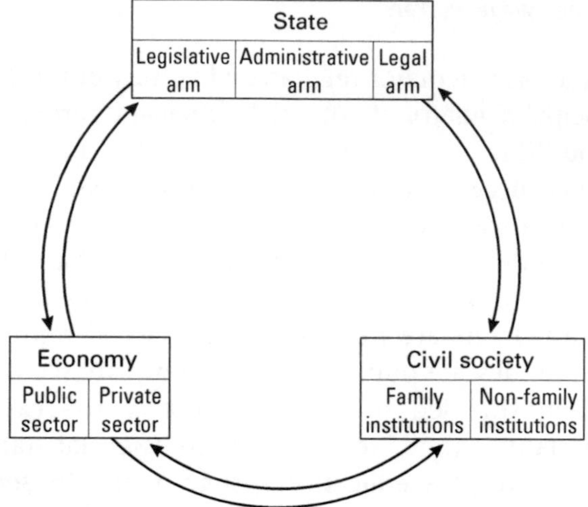

Figure 2.3 The triad of civil society, economy and state

family unit. Though it would be crude to suppose that the family's function in capitalist societies is to engender 'good workers', it is nonetheless a vital institution for producing and reproducing individuals who are able to effectively undertake wage-work. On the other side, the family unit is complemented by a raft of local, national or international charities, non-governmental organizations, foundations, think-tanks, associations and the like that are dedicated to supporting some aspect/s of workers' existence. This plethora of bodies – from housing associations to marriage counselling organizations to churches – together fill in 'regulatory gaps' not plugged by the family and the state (Box 2.4). Of course, most of them were not invented to regulate wage-labour. Rather, all or part of their functions simply happen to lend direct or indirect support to workers in the productive and reproductive spheres.

Regulating dependents, the retired and the unemployed

In the second section of this chapter we explored the intimate connections between waged-workers, their dependents, the retired and the unemployed. As with the two principal players in the employment relation, the latter non-waged groups must be subject to some institutional intervention. Let's take each in turn. Dependents come in two forms: those who are potential future workers (like the children of paid labourers) and those who are permanently dependent

Box 2.4 Non-state institutions supporting worker reproduction

A raft of non-state institutions are involved either directly or indirectly in worker reproduction. They are an important part of what's called 'civil society'. Civil society refers to an intermediate realm situated between the state, the business world and the family, that is populated by organized groups formed voluntarily by members of society to protect or extend their interests, values or identities. Such institutions enjoy some autonomy in their relations with the state. Civil society institutions can be broadly divided into a number of categories:

- Social/recreational organizations: e.g. youth clubs, sports/recreational associations
- Interest-based organizations: e.g. small business associations, professional/occupational associations, trade unions
- Service provision organizations: e.g. marriage counselling, advice bureaux
- Self-help (often community-based) organizations: e.g. urban neighbourhood associations, cooperative societies, credit unions
- Advocacy groups: e.g. environmental associations, women's associations, issue-based pressure groups (representing children, disabled people, immigrants, etc.)
- Cultural/religious/'ethnic' organizations: e.g. religious associations, linguistic groups/literary/cultural associations
- Social movements: organizations dedicated to furthering a specific cause (e.g. debt relief for developing countries) or the interests of a specific social group that lacks formal political representation (e.g. Mexican peasants, women workers)

(like the chronically disabled). In both cases, the state plays a major role in effectively 'subsidizing' both capitalists and workers by taking responsibility for the reproduction of these groups. For example, nursery and school systems help to produce future workers with the 'right' character and skills and are thus a key component of labour supply. Likewise, the state can help workers and their families shoulder the burden of ill, aged or otherwise permanent dependents by offering free or cheap state services for them.

Moving on, the retired, as we observed earlier, are a major social group – especially and increasingly in Western societies where health care improvements today produce unprecedented longevity. Though no longer working, the resources they need to survive require institutional support from state and non-state bodies. Pensions are perhaps the key thing here, with the state either assuming responsibility for pension provision itself or else establishing the framework conditions for adequate private pension schemes. This has a knock-on for business since pensioners are a major consumer group. In other words, in the absence of proper pensions, capitalist firms would be bereft of a major source of consumer demand. We should also note that increasingly many retirees return to work on a temporary or part-time basis. At this point they become workers once more, subject to the regulatory interventions mentioned in the previous sub-section.

Finally, the unemployed cannot be left to their own devices. In a 'pure' capitalist system (what economic sociologist Karl Polanyi (1944: 3) famously called a 'stark utopia'), the unemployed would have to live off whatever monetary and non-monetary resources they'd accumulated during their working lives. For the long-term unemployed, these resources would have to be considerable if they were to survive. In practice, only the most well paid workers have the capacity to accumulate resources of this order. Consequently, without some additional support the unemployed would face insuperable reproduction problems. In turn, this might lead ultimately to social unrest. It has therefore become conventional in most capitalist societies for state and non-state institutions to take some responsibility for both the reproduction costs of unemployment and for the mechanisms to get those out of work back into jobs. Well known examples include the provision of state unemployment benefit, job training schemes, and job centres.

In sum, the social regulation of the employer–employee nexus in capitalist societies is a complex, multi-institutional affair in which the state plays a leading role (Figure 2.4). But it's also an imperfect affair. Social regulation is an art rather than a science. It is prone to hiccups and failures. There are three reasons why. First, social regulation can only manage but never *eliminate* the contradictions of capitalist societies. Second, the sheer number and variety of regulatory objects means that no one body can somehow 'oversee' regulation. Finally, the fact that regulatory institutions must manage people, rather than physical objects, means that there's also likely to be gaps between regulatory aims and outcomes. Thus firms might not *like* the level of company taxation, workers might *disagree* with prevailing employment laws, the unemployed might *protest* over their benefit levels and so on. In extreme cases, the state, as the leading regulatory institution, might face a combination of 'rationality', 'legitimation' and 'fiscal' crises (Habermas, 1976; O'Connor, 1973). The former occurs where the state is unable to keep economic growth on an even footing; the second occurs where intra-economy or economy-wide decline calls the legitimacy of the capitalist way of life into question among

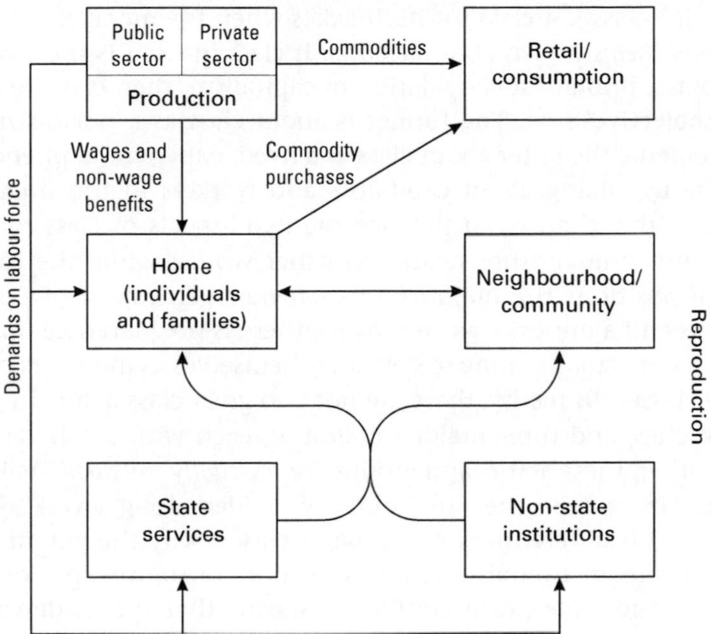

Figure 2.4 The social regulation of labour: an institutional framework (adapted from Gough, 2001, figure 2.7)

large sections of the public; and fiscal crises occur where the state itself, dependent as it is upon tax revenues and investments drawn from the private sector, suffers the knock-on effects of economic crisis. These three crises can occur at a variety of geographical scales and involve local, national or supranational state bodies. They can also occur economy-wide or, less dramatically, within certain specific sectors of the economy, workforce and state.

MULTIPLYING THE ACTORS: THE ROLE OF 'INTERNAL' AND 'NON-CAPITALIST' DIFFERENCES

Regulatory institutions clearly play a vital role in mediating the primary social relationship in capitalist societies – the employer–employee one – along with the secondary relationships with the three other social groupings we've discussed. However, we've tended thus far in this chapter to treat capitalists and workers as what Marxist theorists call 'classes in-themselves' and 'classes for-themselves'. A class in-itself exists by virtue of the *class position* of the individuals who find themselves born into the class (here 'working class' versus 'capitalist class'). A class for-itself exists when the members of it come to *regard themselves as class members.* If class in-itself is the 'objective' side of the primary social relation of capitalism, then class for-itself is the 'subjective' side. The former is about class as a 'socio-structural' phenomena, the latter about class as a lived, experiential phenomena. Despite us talking about capitalists and workers in the singular for much of this chapter, in practice the two aspects of class to do not map onto one another neatly. In other words, while the capitalist system positions the majority of each new generation of people as workers and a minority as employers, there is no guarantee that these people will actually come to consider themselves as members of wider social classes. In reality, there are not two great class actors in capitalist societies, and three major sub-groups, each with a coherent identity. Rather, these actors and groups are *internally fractured.* As Hudson (2001: 218) puts it, they are 'multiply divided along several cleavage planes'. This differentiation is a major reason why the potential class antagonisms of capitalist societies do not constantly spill over into outright and widespread conflict. It means that the co-dependency between employers and workers is expressed differently at the workplace, firm, and sectoral levels rather than finding uniform expression at the level of the economy as a whole.

The cleavage planes involved are at once 'internal' and 'external' to capitalism. We put these two terms in scare quotes because while it is possible to separate them theoretically, in practice they intertwine, complicate and reinforce one another. We've dealt with the former in passing in the preceding pages, but have said little about the latter thus far. Let's take each in turn, devoting most of our attention to divisions within the capitalist and working classes.

Differences 'internal' to capitalism

Social and technical divisions of labour Advanced capitalist economies are made up of millions of workplaces and firms that together constitute different economic sectors. Within these different sectors firms compete for market share. This competitive relation between firms is, in effect, an intra-class relation: it pits both employers and employees within firms against those in others (Hudson, 2001: Chapters 5 and 6). It's thus an *intra-class antagonism.* Thus, despite both being employers in the same industry, the chief executives of Monsanto and Hoescht – two of the world's largest 'life-science' corporations – are in direct competition with one another, as are their workers. Even when firms are not in competition – General Motors, for example, does not target the same markets as Unilever, the pharmaceuticals giant – there remain profound differences between them. This returns us to the social division of labour mentioned in Chapter 1. Complex economies produce *different* commodities for a *range* of intermediate and final consumers by way of *different* combinations of *different* types of labour, technology and materials. Though this extended social division is especially intricate in capitalist economies (because competition compels constant innovation in both production methods and products and also causes the incessant rise and fall of firms and even whole industrial sectors), it is found in all modern economic systems (Sayer, 1995: Chapter 3). It means that both employers and workers exist in relatively autonomous segments of the economy. In other words, if we take any given firm or workplace we find that it has little or no *direct* connection with most other firms in the capitalist world economy. Instead, it will be connected with a specific sub-set (albeit potentially large) of these firms.

For workers, these *intra-class differences* are intensified by what's called the 'technical division of labour'. This refers to the way work tasks are divided up within the workplace and the firm. It was the Scottish economist Adam Smith (1776) who first famously drew attention to this division. He argued in *The Wealth of Nations* that employers could make more productive use of labour if they divided up the labour process into a smaller number of tasks. The wage paid to the worker would reflect the skill level required to complete the task or bundle of tasks. According to Charles Babbage (1835) employers could do away with some occupations by breaking them down into separate tasks. These tasks could then be repackaged into new discrete lower- and higher-skilled jobs. The majority of workers would be the former and would receive lower pay, while there would be less of a need for higher-skilled, and hence well-paid, workers. The 'classic' image of the technical division of labour is the 'Fordist' mass production company where the labour process is broken down into multiple discrete tasks, from research and design to final product assembly (Braverman, 1974). But even small and medium size enterprises – be they farms, local supermarkets or a violin manufacturer – typically involve a horizontal and vertical division of work within and between workplaces. Of particular importance is the separation of the functions of 'conceptualization' from those of 'execution'. In this kind of process, often referred to as deskilling, a task that previously involved thinking a job through, organizing it and then executing it, is broken down so that setting up and performing the task become separate jobs. The net result is the creation of a whole hierarchy of jobs differentiated from each other by their nature and complexity. Some jobs demand more skill than others, some precise types of skill; some jobs are manual, others are mental; some jobs are highly prescribed, others require considerable worker autonomy, and so on. These various jobs are not just quantitatively different but *qualitatively* different. They require workers equipped with quite different bundles of abilities, who then enjoy/suffer a variety of salaries and conditions depending on the status of the job in question. Interestingly, as Massey (1984) describes, the social status of a job may not simply be directly related to the levels of skill and autonomy involved, but is also determined by the social and cultural contexts of the labour market. For example, certain jobs may derive their status from the fact that they tend to be filled by university graduates, rather than any special requirements of the jobs themselves. And the social status of a job is prone to change over time, reflecting and contributing to changes in wider civil society.

Occupations and class fractions Together, the social and technical division of labour within capitalism *fragments* employers and workers both 'in' and 'for' themselves. This fragmentation has knock-on effects in the household, since dependents typically have an interest in their 'breadwinner/s' defending and improving their specific occupations. Sociologists of class have for years been trying to make sense of these so-called 'class fractions' within the wider two-class structure of capitalist societies (see, for example, Wright, 1985). So numerous and shifting are these intra-class differences that it's probably impossible to devise a taxonomy that would adequately capture them. But the point is that however we classify them they *matter,* both objectively and subjectively. In objective terms there is a world of material difference between, say, a low-paid part-time factory cleaner and a McKinsey and Co. management consultant. Not only do they perform vastly different jobs requiring different abilities, but the latter's salary can buy them a lifestyle the former could only dream of. Then, subjectively, it's a well-established sociological fact that people's sense of personal identity is deeply bound-up with their occupations (McDowell, 2003). People in part *define themselves* (and others) by what they do. How they define themselves through their work and what this means for them is shaped by issues of gender, ethnicity, and so on (Pollert, 1981). Moreover, this has social status implications. Thus, in a given workplace, managers might consider themselves to be of a 'higher' social rank than those they manage – despite their shared situation as employees. Likewise, the owner of a small 'clean technology' firm – producing solar panels, say – might consider his/her occupation more 'respectable' than that of the CEO of a 'dirty technology' firm (like a coal and oil company).

All of this does not mean that employers and employees are isolated at the workplace or firm level. On the contrary, it is in their interest to make formal links with others in the technical and social division of labour. For employers, as we noted earlier, industry associations can help to lobby the state on behalf of groups of firms within a specific sector. The major European life-science companies, for example, have voluntarily formed EuropaBio, an organization that represents their economic interests at the global level (at World Trade Organization meetings, for example), at the European level (in the European Union legislature) and the national level (in national state legislatures). Likewise, trades unions help to organize workers within and between firms, building solidarity among workers otherwise differentiated by occupation, workplace, firm and industry. But while these solidarity-building institutions link employers and employees respectively, they can also entrench existing divisions. For example, many trades unions are organized on an industry basis. Take the United Steelworkers of America (USWA). As its name suggests, it is the principal union for US steelworkers, whichever steel company they work for. But it thereby cordons these workers off, at one level, not only from other workers in heavy industry but labourers in other economic sectors. Similarly, within firms and economic sectors trades unions separate workers by occupation. For example, although we authors work in the same institution as Nick Scarle, the cartographer who produced many of the figures in this book, we belong to a different union from him.

Other axes of difference Finally, there's two other important sets of cleavage plains that are nominally internal to capitalism: those dividing the employed, the unemployed and the retired, and those dividing workers on the basis of the permanence, periodicity, legality and site of their jobs. Let's take each in turn. Earlier in the chapter we stressed the connections between the employed, unemployed and retirees. All three depend upon, or aspire to, wage-work for their long-term survival. On a less abstract level, wage-workers may actually live with relatives seeking a job (and thus seeking to boost household income) or drawing a pension. However, in practice there are some tensions and differences between the three groups. The unemployed, of course, are potential competitors with the employed: competitors, that is, for jobs. This can become acute during times of both high and low unemployment within the economy. In the former case, for example, the surplus of potential workers can enable employers to freeze or reduce wages among existing employees. Conversely, when employment levels are high, it can be very difficult for the remaining unemployed to find work. In practice, this competition is generally highest over jobs found at the lower end of the labour market where there are fewer 'barriers to entry' for prospective workers. Meanwhile, the retired simply may not realize that their pensions at some level depend upon the aggregate retirement plan contributions of present day employees.

As if the various cleavage plains internal to capitalism were not enough, within the labour-force we must recall that workers' objective conditions and subjective experiences vary according to whether they're permanent or temporary, full- or part-time, legal or illegal, and home- or workplace workers (see Figure 1.2 again). It's often very difficult for workers to identify what they have in common when one is, say, a seamstress doing part-time piecework at home for a clothes contractor while another is a Harvard educated full-time games console designer for Sony. Those on the fringes of the labour market – that is, those doing temporary, illegal, part-time work – tend to move in and out of the unemployed category, thus accentuating their differences from those in regular, permanent employment.

Differences 'external' to capitalism

Capitalism does not exist in a social vacuum. In both its objective and subjective dimensions, class is deeply intertwined with several non-capitalist axes of social differentiation. These axes of differentiation, as Peck (1996: 31) argues, 'are not *created* by capital[ists or workers, but] are nevertheless [influential]' in employer–employee relations. The principal ones are gender, sexual preference, ethnicity, nationality, age and religious belief. They have three things in common. First, like class, they are *social* axes of differentiation: that is, they precede and transcend the specific individuals who are 'slotted' (or slot themselves) into them. Second, they are *ascribed* axes: that is, they are not 'given' in a person's character or behaviour. Thus all women share certain obvious biological similarities. But these bear little relationship to gender, that is the socially specific ways in which what it means to be a 'woman' are defined. Finally, these axes serve as both *resources for and restraints upon* individuals. For instance, in a heterosexual society, someone who 'comes out' as gay may experience a feeling of release and liberation at being able to call themselves a homosexual. But, at the same time, in many societies homosexuals suffer discrimination because they contravene the norms of heterosexuality. The precise way and the exact degree to which gender, sexual preference, ethnicity, nationality, age and religious belief have stigmas or kudos attached to them clearly varies between societies. And, equally clearly, these axes of difference do not magically vanish when a person seeks work. Rather, individuals are 'marked' by sets of ascribed characteristics that affect both the kind of jobs they seek and the kind of jobs they are likely to be offered.

The specific way that class and non-class differences articulate is a complex matter. A given individual might be marked by all the 'non-capitalist' differences that, in their society, render them vulnerable to labour market or workplace discrimination. Another person might, conversely, possess the 'right' cluster of ascribed characteristics. In any particular case we need to consider how ascribed categories of difference are both perceived by others (workers and employers) and by those who are marked by them. If we take the former, what's striking in all societies worldwide is that certain groups of individuals seem, time and again, to 'win' or 'lose' in both the labour market and the workplace. Men, of course, are typically offered better jobs, pay and conditions than women! In predominantly Caucasian countries, people of colour seem unable to access jobs high-up the occupational ladder, while whites enjoy better occupational prospects. For instance, in the UK Pakistanis are twice as likely to be unemployed as Indians are (Cabinet Office, 2002). Meanwhile, in much of the Muslim world, non-Muslims rarely ascend to the upper echelons of the labour market. We could go on, but the point is hopefully clear: potential and actual workers are ranked by employers (and their intermediaries) and matched to certain types and conditions of work.

Turning, finally, to how individuals internalize the social meanings attached to their ascribed characteristics, many of them may come to think of themselves as class individuals second, and men or women, black or white, gay or straight, etc. first. So non-capitalist axes of difference deeply affect workers' self-identity – perhaps to the point that they have a very weakly developed class consciousness. In Hudson's (2001: 106) words, 'While they share a commonality of interest as wage workers, they do so bearing other identities that potentially divide them'. Class solidarity can, therefore, be diluted or challenged by non-class bases of identity and affinity.

SEGMENTED LABOUR MARKETS, SEGMENTED WORKERS

When we add together the raft of regulatory institutions, regulatory mechanisms, intra-capitalist divisions and non-capitalist differences discussed in the last two sections we start to appreciate the extraordinary complexity of the social landscape of wage-labour in the modern world. Though the logic and relations of capitalism may lie at the heart of things, they are so internally riven and so influenced by nominally non-capitalist institutions and processes that the two-class model with which we began this chapter starts to look implausible. In reality, we have a fractured capitalist class and a fractured working population (with dependents), while the unemployed and retired are linked to but not necessarily allied with the latter. None of this alters the fact that all working people find themselves in a common class position (class in-itself). But it helps to explain why capitalist societies do not constantly erupt in widespread conflict between the capitalists and workers. Furthermore, it indicates that if a common class identity (class for-itself) is to be developed among workers it requires a considerable *effort* of stressing what unites them rather than what divides them. A wider class consciousness does not come 'naturally'; instead, it has to be constructed. In this penultimate section, we bring the insights of the previous two together in order to see how there's a degree of *order* to the *heterogeneity* among capitalist wage-workers – what Tilly and Tilly (1998: 256) aptly call 'principles of variation'. Though multiple, the differences between workers are not random. In metaphorical terms, the social landscape of labour has numerous but nonetheless relatively enduring contours.

These contours are all about what labour market theorists call *segmentation*. As one so-called 'segmentation theorist' has put it, 'What distinguishes segmentation from mere division is that each segment functions according to different rules' (Michon, 1987: 25). In essence, different individuals and groups sell and deploy their labour power in various 'sub-worlds of work' within the wider universe of capitalism. Movement between these worlds can be quite difficult for workers, who tend to be channelled into different labour market and workplace segments. This segmentation process is the combined effect of three things (Peck, 1995: Chapter 3; Storper and Walker, 1989: Chapter 6): specifically, segmentation of labour supply, labour demand and of workplace activities.

- **Segmentation of labour demand:** there is enormous variability in the quantity and quality of labour power sought by different firms. The technical specifics of each productive process generate fine-grained variations in demand for varying numbers of workers with specific characteristics. In all cases, as we've seen, employers seek to balance their desire to exert workplace control with the need for workers to cooperate. This generates 'labour queues' (Thurow, 1975). Here, employers rank different groups of workers coming to the labour market as to their likely 'goodness of fit' with specific occupations. In Storper and Walker's (1989: 167) words, 'Lacking information on individual workers, they must use indirect indicators. ... [E]mployers try to identify background characteristics that are good predictors: educational level, for example, may indicate an ability to absorb training ...'. Not surprisingly, when ranking potential workers in this way, employers resort to using conventional categories of social difference as proxies for any individual's suitability for a particular job. It's no surprise that those individuals whose gender, sexual preference, ethnicity, religion, etc. make them vulnerable to social discrimination also face labour market discrimination.

- **Segmentation in workplaces:** of course, having been ranked by employers even before securing employment, once in the workplace labourers may find it even harder to break out of the occupational segment to which they've been assigned. Many occupations do not have 'career ladders' attached to them: that is, possibilities for promotion and progression. Other jobs require such low or specific skills that those who undertake them find it hard to move into different or better occupations. This form of segmentation results from the combined effects of technical and social divisions of labour. Because of the combined effects of pre- and in-work segmentation, employers are thus able to place metaphorical 'walls' between workers.
- **Segmentation of labour supply:** as noted earlier, the supply and reproduction of workers is largely beyond the control of firms. But the combined effect of the family, the education and training system, the various institutions of civil society, the conventional non-class axes of social difference, and so on is to sharply segment the supply of labour. Thus, in those many instances where men expect women to undertake the lion's share of domestic work and childrearing, those women are unlikely to take on full-time wage work, however much they might want to. Somewhat differently, because children internalize the values and aspirations of their parents, they frequently perpetuate them. Thus, as sociologist Paul Willis (1977) famously showed in *Learning to Labour*, the reason lower-working-class children usually end up doing lower-working-class jobs is because – like their parents – they were unwilling and unable to access educational opportunities that would break the cycle of low pay and low aspirations.

When combined, the segmentation of labour demand, labour supply and workplace activities allow us to make sense of the fault-lines within the labour force (Figure 2.5). These fault lines are arguably *systematic,* rather than shifting, within and among countries worldwide. We can thus talk about *structured differences* in job type, occupational mobility, promotion prospects, salaries, working conditions

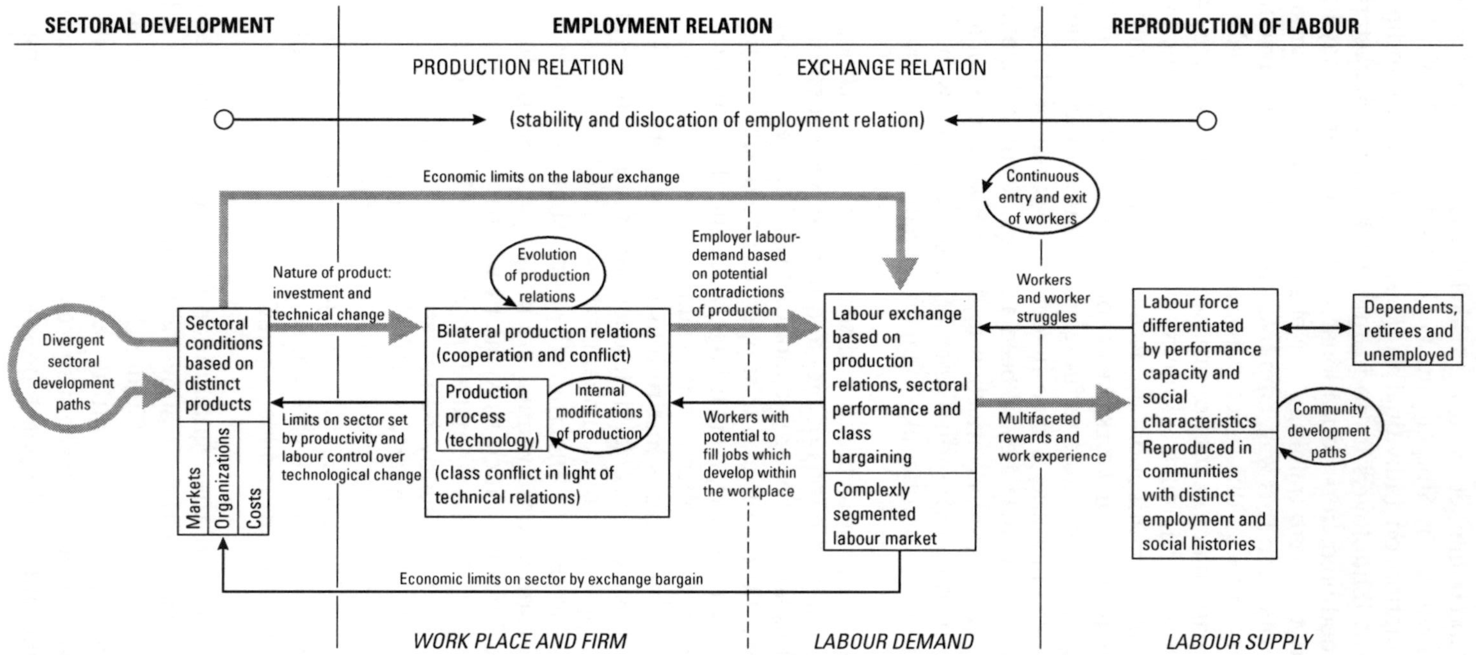

Figure 2.5 The segmentation of labour (adapted from Storper and Walker, 1989: figure 6.3)

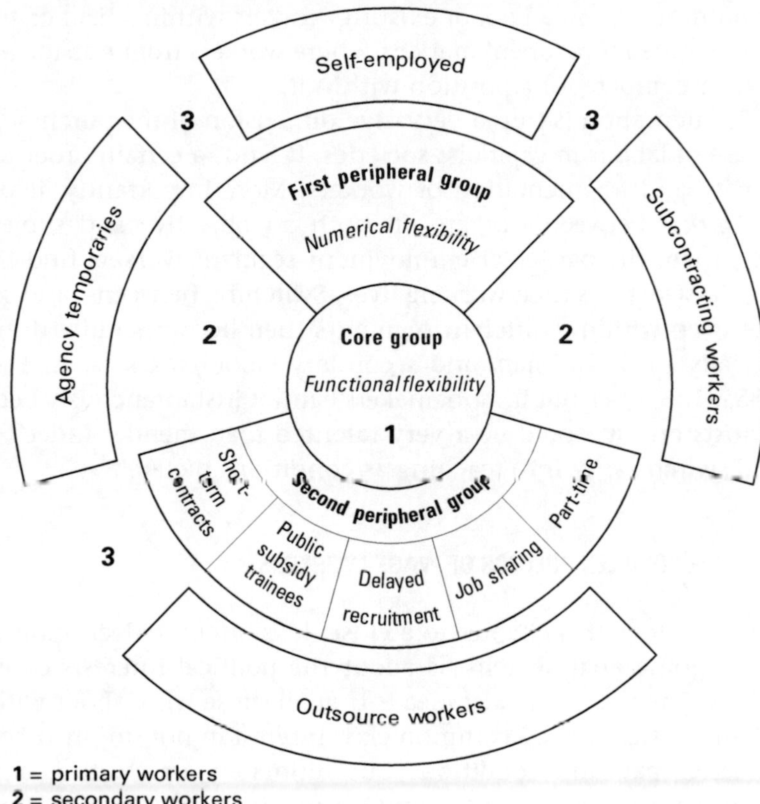

Figure 2.6 Labour segments under capitalism (adapted from Peck, 1996: figure 3.1)

and non-salary benefits for different groups of workers. To hazard a generalization, we can distinguish 'primary', 'secondary' and 'tertiary' labour markets, each with their own distinctive worker groupings. The former involve 'core jobs' that are full-time, well-paid, self-directed and relatively secure, requiring well- or highly-educated and skilled workers; secondary labour markets involve full-time, medium-to-poorly paid, moderately self-directed and relatively secure jobs, requiring only moderately educated and skilled workers; finally, tertiary labour markets involve temporary or otherwise irregular jobs that are poorly paid, prescribed and very insecure, requiring workers who are poorly educated, unskilled, and socially marginal. Each of the three labour markets can, in turn, be divided into submarkets (Figure 2.6). Additionally, across the trio we can identify *internal* and *external* labour markets. The former are 'sheltered' markets where workers are hired and promoted from a pool of existing workers within a firm or sector. The latter are more 'open' markets where workers from outside a firm or sector compete for a position within it.

Segmentation is thus a pervasive dimension of the purchase, sale and use of labour in capitalist societies. It lends a certain order to the diversity and fragmentation of workers. More importantly, it places real *barriers* between workers. In both an objective and subjective sense, then, the particular employment segment workers find themselves in structures their working lives. Switching between – and sometimes even within – different segments then becomes quite difficult, especially in the primary and secondary labour markets. As Harvey (1985a: 136) once put it, 'shoemakers cannot instantaneously become scientists and it would be a very talented road mender indeed who could switch easily into teaching as conditions dictate'.

THE COMPLEX POLITICS OF WAGE-WORKERS

Before we draw this chapter to a close, let's briefly reflect upon what the foregoing analysis tells us about the political interests of wage-workers. The classic Marxist idea is that all those who labour within a capitalist system have a common class interest in not simply demanding better pay and conditions but ultimately in creating a post-capitalist system where the many do not have to work for the few. However, what we've seen in this chapter is that the political interests of wage-workers are necessarily more complex than this because of the multiple social divisions in play. Although somewhat simplistic, the following typology can help us get a handle on the kinds of political agendas that different workers might hold or that the same workers might expound simultaneously:

- **Class politics:** here, as in the Marxist view, a worker might stress his/her interest in achieving better pay and working conditions as part of a wider working class or class fraction.
- **Occupational politics:** here the class element of a worker's existence is less important than their occupation, which becomes the principal way they define themselves as a worker (for example, a website designer; a teacher).
- **Identity politics:** here a worker might choose to stress his/her interest in achieving better pay and working conditions but not in class or occupational terms. Rather, the worker will emphasize a relevant non-class aspect of his/her identity that seems especially relevant to their working lives. Thus, females might in certain cases

emphasize their shared identity as women in order to highlight how they typically receive lower pay and less esteemed jobs than men. Lesbians, gays, people of colour, people of minority religions and so on can all lay claim to some aspect of their identity in order to show (and contest) how they suffer workplace discrimination because of it.

- **Reproduction politics:** this is directly related to but not the same as class politics. Here the unemployed and the underemployed might seek the simple right to work and receive a salary for themselves and their dependents. This, then, is a politics of the right to reproduce: to receive a living wage, to ensure physical and some degree of social reproduction.
- **Life politics:** this involves workers coming together with all manner of groups in civil society over 'quality of life' issues that include but extend beyond workplace concerns. Thus, the right to live in a clean environment is a 'life politics' issue that affects *all* people – workers and non-workers. It is important to note that life, reproduction and identity politics are not *reducible* to issues of work but, rather, impinge upon them.

Even with a simple five-part scheme like this one, it's clear that how any given worker chooses to define his/her main work-related interests is complex and potentially variable. Imagine a wealthy, Aboriginal, female, newspaper editor in Australia. Where do her working interests lie? Is she primarily a wage-worker who should, therefore, take a direct interest in those less well paid then herself? Is she, instead, essentially an editor (an occupational category) or, rather, a woman in a world where few women ever reach the upper reaches of the occupational ladder? Perhaps, to raise one other possibility, she might choose to stress her Aboriginal identity and thus ally herself to all those Aboriginals in Australia much less fortunate than herself. In reality, of course, her interests lie in all these domains simultaneously. Which aspect/s of her working existence she chooses to emphasise in any given instance, and which other workers she might ally herself with, is an open question, but one that informs the politics she might engage in.

SUMMARY AND PROSPECT

In this chapter we have explored the nexus of social relations and institutions in which paid labourers are immersed. We've seen what a complex nexus it is, one that belies the simple notion that workers, employers and ancillary groups have coherent and consistent interests and identities. Wage-workers are socially connected *and* differentiated, socially similar *and* different, socially linked *and* dissociated. To understand how they, their employers and non-waged social groups interact in any given case is thus a contingent matter. It depends on how regulation, internal difference, external difference and axes of commonality *combine* for individuals, families, households, workplaces, firms and industries. The nature and potency of the power repertoires specific employers and employees can bring to bear in a given situation is thus dependent on who they are and the precise regulatory, and economic conditions in which they find themselves. But it also rides, in quite profound ways, on geography. This is the subject of the next two chapters. We have abstracted the social dimensions of wage labourers' existence from their geographical integument. It's now time to explain why geography makes such a difference to what is currently happening to workers worldwide and how they respond to their circumstances.

three

Placing Labour in an Interdependent World

Geography matters to workers. Workers, conversely, matter to geography. But what do we mean by 'geography'? And in what senses does it 'matter' to workers (and vice versa)? These are questions we address in this and the following chapter. As we'll see, geography is anything but an 'optional extra' to our analysis. Instead, we treat it as what Harvey (1985b: xi) calls 'an active element'. At first sight, this might strike you as a peculiar claim. After all, students of sociology, economics, politics and the like often think that geographers simply catalogue endless facts and figures about the earth. Meanwhile, those doing degrees in human geography – who should know better – frequently fail to see the theoretical significance of the things they study. For too many years, human geography was the 'poor relation' of the social sciences. It was the discipline that took the theories of other subjects and showed how the processes depicted in them worked out differently in different places. However, in the last decade, it's become clear that what human geographers study is of more than merely empirical significance. As this and the following chapter will demonstrate, this is manifestly true of wage-labourers seeking to make a living in a capitalist world. If theorizing is about identifying the fundamental processes at work, then geography has theoretical significance

(Massey, 1985). It simply cannot be divorced conceptually from the social issues discussed in the previous chapter. Nor can it be treated merely as the domain where social processes find 'empirical expression' on the ground. Instead, it is *constitutive* of those processes. More particularly, we argue that *place, space* and *geographical scale* are the concepts we need to grasp if the current condition and future prospects of capitalist workers are to be properly understood. We start, in this chapter, by showing how wage-workers' existence in a capitalist world is still fundamentally local in nature (place-based). In the next chapter, we will introduce the concepts of space and scale to develop our geographical framework. Though the structure of Part One separates out the non-geographical and geographical elements that constrain and enable the lives of wage-workers, this is purely for analytical purposes. In the real world, these elements are intimately intertwined.

WHAT MAKES A PLACE?

Objective and subjective dimensions

We begin by considering both the nature and the enduring ontological importance of place. By place we mean the scale of everyday life: the scale of cities, towns and communities. Geographer Tim Cresswell (1999: 226) has observed that 'place is ... a term which eludes easy definition' (see Staehli, 2003). At its simplest, it means two familiar things: first, a distinct point on the earth's surface; and second, the local scale (hence the interchangeability of the terms 'local' and 'place' in the previous paragraph). The linked notion of *locality* captures this double-sense of how people, institutions and things 'come together' in unique ways in different locations to create a plethora of places that are more-or-less distant from one another. Places are not given in nature. Rather, they are socially constructed. Though conditioned by local environmental circumstances, places are the product of myriad human practices. It's the sedimentation of these practices over time that lends these places their distinctiveness.

Even in the supposedly 'frictionless' world of the new millennium, what Cox and Mair (1989) once called 'locality dependence' looms large. This term describes social relationships, regulations and institutions that have a very high degree of local embeddedness. They are either exclusively local or partly local. As Cox (1998: 2) puts it, 'people depend [upon them] ... for the realization of essential interests and ... there is no substitute [for them] elsewhere; they define place-specific conditions for [people's] material well-being.' Locality dependence means that the principal social relationship of capitalist societies, that of class, is really *a set of place-specific class relationships* between firms and labourers, mediated by regulatory institutions. In keeping with our discussion of class in Chapter 2, these relationships involve different sets of 'internal' and 'external' axes of social difference among workers and employers coming together in unique ways in different places. Places are thus complex ensembles of workers, dependents, retirees, the unemployed, firms and institutions that are built-up and variable over time. They possess what Harvey (1985b: 139) has called a 'structured coherence'. Though composed of multiple people, businesses, organizations and built environments, there's a certain order to this place complexity. Places take shape through what sociologists call the 'routinization' of the relationships between local actors and institutions. These actors and institutions must be able to ensure some regularity to their interactions. Otherwise, daily life would be chaotic or anarchic: neither production, reproduction nor consumption – the lifeblood of existence – would be possible.

Relatedly, we must also recognize that this patterning of everyday activities in place has a profoundly subjective and affective dimension. Places are where people – workers *and* employers, the unemployed *and* the retired, dependents young *and* old – *live*. It is sometimes said by Marxist theorists that people living in places are primarily interested in the 'use value' of these places, while theorists of a more hermeneutic bent emphasize their emotional value. Use values are all the practical benefits that the infrastructure of a place yields, while emotional values are the equally real, if less tangible, benefits deriving from people's attachment to where they live. Place gives meaning to people, and people give meanings to places. If they didn't develop attachments to place or if they didn't care about what happens to their home place, it would be as if people lived and worked on an isotropic plain. Accordingly, different places possess their own *genre de vie*, or what cultural critic Raymond Williams (1975) once evocatively called 'structures of feeling'. These affective ties can unite some or all local actors and institutions in *joint* commitments to the place where they commonly reside (as we shall see in Chapter 6).

Conceptualizing place

So place, in both an objective and subjective sense, remains of ontological importance in the contemporary world. Places are both material arenas for the conduct of everyday life and loci for the development of local identities and loyalties. But this raises a key question: namely, how should we *conceptualize* places in the twenty-first century? It's often thought that what makes a place is what goes on *within* it. However, there are two problems with this common-sense conception of place. First, it implies that there are distinct boundaries between the 'local' and the 'non-local'. This is what Phil Crang (1999: 27), in a review of place concepts, calls a 'mosaic view' that is desperately outdated. From this perspective, places are like discrete pieces of a larger geographical mosaic, such that the 'insides' and 'outsides' of places are readily apparent. Second, it is increasingly obvious that we live in a world where places are not only interconnected but also *interdependent*. This is one of the things the term 'globalization' has sought to capture. The implication is that what makes a place has everything to do with 'distanciated' events, processes and institutions stretched-out over a larger *space*. It is no longer correct to assume that what is geographically juxtaposed is more important than what is geographically separated. Instead, geographical presence and absence co-mingle in varying ways.

If we take these two points together, we begin to see places less as bounded areas within a larger space of interconnections and interdependencies, but as the *meeting-place* of bundles of 'local' and 'non-local' events, processes and institutions. The geographer Doreen Massey has arguably done most to popularize this *relational* view of place. In her words, we now need to conceptualize places as 'particular articulations of ... social relations, including local relations 'within' the place and those many connections which stretch beyond it. And all of these [are] embedded in complex, layered histories. This is place as open, porous, hybrid ...' (Massey, 1999: 22). Here, then, space and place are seen to melt into one another. Though one can distinguish them theoretically, in practice they interfuse. To quote Massey (1995a: 54) once more: 'social relations have become so stretched out ... that it is difficult any more to distinguish within social space any coherent areas which might be called places ...'

This conception of places as permeable has three of important implications and advantages. First, it shows us that it's misguided to look solely 'within' a place if we're to understand what locally situated workers, firms and institutions are up to and what the effects of their actions are. A simple way to illustrate this is to consider the activity spaces of any given worker (Figure 3.1). For example, in our case, we all live within 12 km of our place of employment. But our production, reproduction and consumption activities – in the short

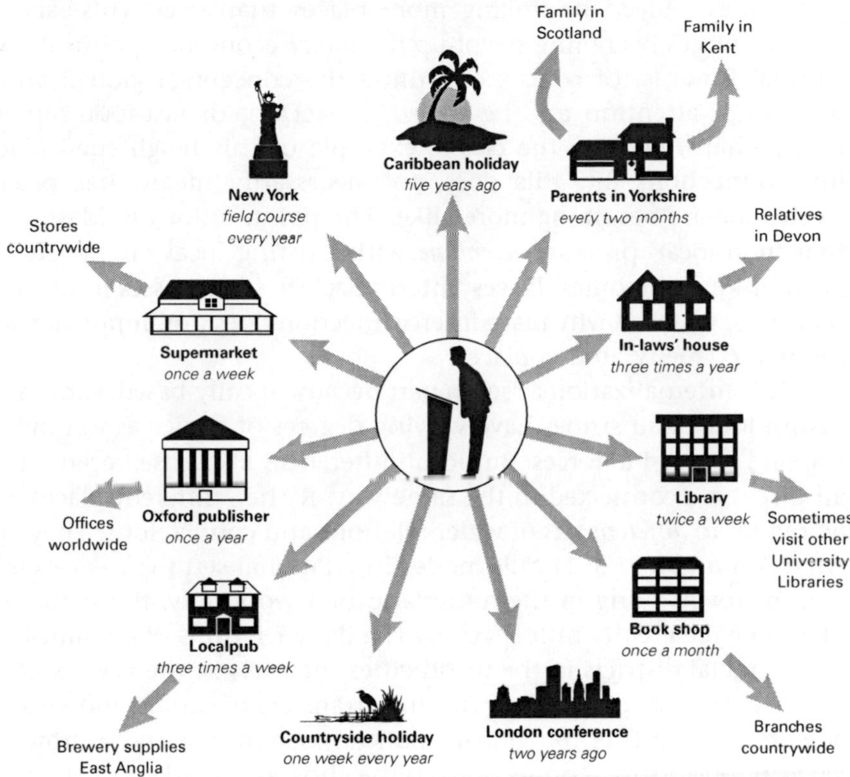

Figure 3.1 A university professor's immediate activity space

and long term – implicate us in a far-flung geography that easily escapes the confines of the office in which these words are being typed. So places are increasingly not just local: they are what Marxist geographer Erik Swyngedouw (1989) memorably called 'glocal'. By coining this neologism, Swyngedouw seeks to stress that the non-local is increasingly *in* the local.

Second, though, this first point leads to an apparent paradox. If what makes places is not simply internal to them, and if we're arguing that places are *unique* combinations of people, institutions and firms, then how can uniqueness persist when places are increasingly influenced by 'outside' forces? In other words, how can place differences persist not *despite*, but *because of*, heightened place interconnections? – which is what a relational view of place suggests is happening. The answer is as follows. Translocal interconnections, whereby commodities, people, information and images criss-cross the globe, may indeed be linking more places than ever. This fact of *different* places becoming swept-up in *similar* economic, political and cultural flows is, of course, one thing the concept of globalization directs our attention to. The spread of McDonalds fast-food outlets has, perhaps, become the classic example of this heightened place interconnection. But this does not necessarily mean that places worldwide are becoming more alike. The point, following Massey, is that 'non-local' processes *combine* with existing local differences to yield unique outcomes. Places 'internalize' these processes in distinctive ways, which is why place interconnection does not imply increasing homogeneity among places.

This internalization arises in part because locally-based actors and institutions of all stripes have varying degrees of choice as to how to respond to outside forces. But local differences also arise because not all places are connected in the same way. Rather, different places are 'wired-in' to *different sets* of wider relations and connections. They are what Amin (2002: 391) calls 'nodes in relational settings'. For example, for towns lying in the hinterland of a world city, the nature of place may be greatly influenced by the daily rhythms of commuting. For financial districts in the world cities themselves, the very essence of place depends on the long distance transfers of capital and knowledge that constitute the global financial system. For cocoa growing areas in West Africa, connections to the chocolate markets of Western Europe are critical, and so on. Of course, in reality, a complex intermingling of myriad local and extra-local connections determine the nature of most places.

Third and finally, seeing places as open and porous allows us to understand why they are not simply different from one another but *unevenly and causally related*. This brings us back to an important distinction made above between place interconnection and place interdependency. The former term merely states that places are somehow interrelated in the early twenty-first century. The latter term, however, is more precise and meaningful: it implies that the very *nature and fortunes* of places are bound up with what is happening elsewhere. Think again of the Malmesbury-Malaysia example involving Dyson, mentioned in Chapter 1. It's precisely the place differences that have impelled Dyson to abandon production in the former locality. But the two places are directly co-dependent because new investment in the latter spells severe job losses in Malmesbury, despite them being thousands of kilometres apart. So, there's an element of mutual determination in place relationships, which both stems from and alters existing place differences. As John Allen and Chris Hamnett (1995: 235) have put it, 'the networks of social relations stretched across space are not simply uneven in their [geographical] ... reach, they also *work through* [place] diversity and difference. Geography matters in this instance, precisely because [place] relations construct unevenness in their wake *and* operate through the pattern of uneven development laid down' (first emphasis added). Uneven geographical development – in terms of everything from the average cost and standard of living to the average skill level of workers – is thus both a precondition and a cause of place interdependency in the contemporary world. Place inter-dependence, in other words, produces 'winners' and 'losers'.

In sum, twenty-first-century places tend to be open and interdependent, yet without losing any of their individuality and uniqueness. The fate of places is thus only partly in the hands of those who inhabit them. Though the word 'place' traditionally conjures up images of a relatively fixed and settled community, it's increasingly clear that in a capitalist world places are dynamic and ever-changing. To adapt a phrase from Marx and Engels' *Communist Manifesto* (1853), the solidity of places constantly melts into air. The once prosperous steel-making town can thus become today's unemployment blackspot, while yesterday's sleepy university city can become tomorrow's hi-tech growth pole. Quite simply, the rise and fall of places is a fact of capitalist life.

PLACING WORKERS

Why place matters to workers

Everyday life in a capitalist world, we are arguing, is simultaneously local (placed) and translocal (spaced). In this and the subsequent sections of the chapter we take each domain in turn. We begin by explaining something that was merely asserted in the previous section: namely, that wage-work and wage-workers are *necessarily local*. In our view, place forms the immediate objective and subjective arena in which all virtually *all* paid work is undertaken today. As two of the foremost geographical analysts of workers have put it, 'Although the world is increasingly well-connected, we must hold this in balance with the observation that most people lead intensely local lives: their homes, workplaces, recreation, shopping, friends and other family are all located within a relatively small orbit' (Pratt and Hanson, 1994: 25). Why, then, in our supposedly globalized world, are workers unavoidably *place-based*? It seems to us that there are seven reasons:

- First, most workers have neither the time nor the resources to travel far to where they work. Even in the advanced capitalist economies, the average person travels just 25 km per day (Holloway and Hubbard, 2001: 27), while elsewhere in the world the figure is nearer 5 km. As David Harvey (1989: 19) once put it with disarming simplicity, 'Labour-power has to go home every night.'
- Second, virtually all production activities are necessarily local. Only 'mobile' businesses – like air travel – involve labour 'working on the move'. For the vast majority of paid workers, production occurs at fixed sites. This is true even for the employees of the largest TNCs, like Microsoft or General Motors. Though these firms are *relatively* more mobile than most small and medium sized businesses, they must still combine labour, materials and technology in work*places*. Without this physical anchor, commodity production would simply be impossible.
- Third, the reproduction of labour is also necessarily local. To borrow Storper and Walker's (1989: 157) felicitous words,

> It takes time and spatial propinquity for the central institutions of daily life – family, church, clubs, schools, sports teams, local unions, etc. – to take shape ... Once established, these outlive individual participants to benefit, and be sustained by, generations of workers. The result is a fabric of distinctive, lasting local communities ... woven into the landscape of labour.

- Fourth, as part of reproduction, we noted in the last chapter that workers are not just commodity producers but commodity consumers. Without the purchase, use and subsequent purchase of commodities capitalism would grind to a halt. This has a place dimension because consumption is necessarily local for most people most of the time. A physical location is required to both sell and consume commodities, however far away those commodities may have been made and transported. Moreover, certain consumption practices take on a distinctly local character. For instance, eating sushi has become a new lifestyle favourite among urban professionals in English-speaking countries, while owning a BMW or Mercedes car in Singapore is a marker of real distinction.
- In the fifth place, for the four reasons stated above, the regulation of workers is locally expressed. Even where the regulatory institutions are national or international, the fact of workers' local existence means these institutions' regulatory mechanisms must ultimately be articulated at the local scale. For example, most

trades unions are principally national organizations. However, they are nonetheless constituted by myriad local memberships that, together, make-up the national union. In turn, any actions taken by the national union leaders on the local memberships' behalf must ultimately be agreed to and acted upon by this membership (we take up the issue of non-local scales later in the chapter).

- Sixth, in light of the above it's plain to see why labour markets are *necessarily* local. The sale and purchase of wage-workers occurs in conditions where labour and employers are place-based, while regulatory institutions must find local expression if they are to influence the employer – employee nexus. Since the exact bundle of workers, employers and institutions is unlikely to be the same from place to place, we can thus say that labour markets are locally *constituted* (Peck, 1996: Chapter 4): they '*operate* in different ways in different places' (ibid: 86).
- Finally, because labour is not like other commodities it has the capacity to develop place attachments. The most articulated form of these attachments are *place identities*. The term 'home place' connotes something of this insistently local dimension to many people's sense of themselves. Places become the locus of familiarity, affection and even love. The well-known sites and smells, the daily routines, the ever-present landmarks, the enduring friendships, the family ties: even today, these remain overwhelmingly local. What's more, this sense of being *from* a place – of place somehow being *part of oneself* – is hardly exclusive to workers. Employers too develop deep place attachments, for they must also live somewhere. Likewise, those who do not undertake paid work – especially children and the elderly – are often confined to highly localized activity spaces because their mobility is restricted. These local (and sub-local) spaces become the primary 'experiential contexts' of these groups.

In short, there is a certain 'stickiness' to workers, even at a time when the mobility of people, business and commodities is seemingly at an all time high. This said, labour migrants clearly have *routes* rather than simply roots. That is, they voluntarily or involuntarily move over wider spaces in order to access employment opportunities. As we will see in Chapter 7 this mobility – however unwanted or stressful it might be for the workers involved – can be seen as a way to escape the confines of labour's placed-based existence. Indeed, labour migrants show that while wage-workers are typically place-based they are not necessarily *place-bound*. However, this fact notwithstanding, even migrants cannot ultimately escape the pull of place. In simple terms, they always migrate *to* a place having travelled *from* another. Moreover, if they are supporting dependents back in their home place then this exerts an enduring, material influence on their daily lives, even in their new place of work.

To summarize, despite all the hype about living in a 'global village', wage-workers exist in multiple 'local worlds' that are geographically separated (though not isolated). Even in the twenty-first century, the daily journey to work remains local, the reproduction of workers remains place-based, commodity production and consumption remains locally embedded, and the regulation of workers and employers is locally expressed. Meanwhile, the character of places is written into the identities of those who are born and raised in them, and sometimes into the identities of those who move permanently to new ones.

Workers within place

What the previous sub-section shows is that wage-workers are *geographically differentiated* among localities. They live, learn and labour in different places within multiple nation-states worldwide. This geographical differentiation adds a further layer of division and complexity to the social differences between workers explored in the latter part of the previous chapter. It's what we call the *spatial division of labour* (as mentioned on p. 8): that is, the way the social division of labour is expressed territorially. But we can take things even further. Though we implied it above, it is wrong to think that different places possess relatively homogenous – or at least similar – groups of wage-workers in each of them. True, so-called 'single industry' communities or 'company towns' do have especially non-diverse working populations. The classic examples are the former coal mining communities of Appalachia or Northern England. But many – indeed, perhaps most – places consist of *several* groups of workers who labour, reproduce and consume under rather different conditions. In short, we need to appreciate the divisions between workers *within place*, even when those workers share a common attachment to the place in question. If you like, there are 'sub-local worlds' of existence within the 'local world' of a given place.

The fault lines of labour market segmentation described towards the end of Chapter 2 clearly run through, as well as between, different places. There are important differences between workers within places regarding attributes such as skills, gender, ethnicity and age, but also in terms of their attachment to place, seen both in socio-cultural

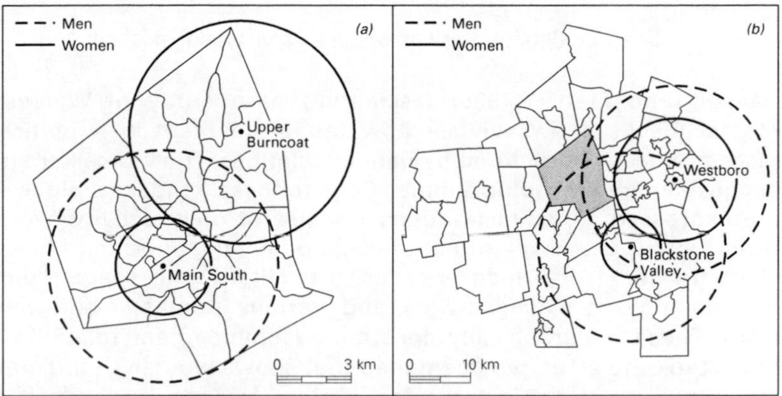

Figure 3.2 Median journey-to-work distances of men and women in four local areas in Worcester, Massachusetts: (a) city of Worcester, (b) Worcester MSA

Source: Hanson and Pratt, 1995: figure 4.2

terms and also through different levels of mobility within the housing market. Sub-local segmentation is often at once both occupational and *geographical*. Certain groups can become 'geographically entrapped' within certain districts. Unskilled low income workers, for example, are more likely to change their job than their residential location. They are also less able to bear the costs of daily travel to distant districts, limiting the range of employment opportunities available to them. Where no suitable employment opportunities are in reach, pockets of unemployment may develop. Women in many places share some of these difficulties due to their need to balance paid work and non-paid domestic and caring responsibilities (McDowell, 2001) (Figure 3.2 and Box 3.1). On the other hand, highly paid workers not only have the financial means to travel greater distances, but also to shift residential location. In some instances this will involve relocating to another locality. Places, then, are internally 'balkanized' with workers differentiated in terms of where they live, where they work, and their potential mobility between the two arenas. The nature of the housing systems often exacerbates balkanization both in terms of where particular types of housing are situated, and the mechanisms through which the stock is allocated. One only has to think of a typical city and the differences between the inner city and suburbs to see these processes of segregation in action. In some instances, such is the effectiveness of this social and sub-local segmentation that different groups may never – or only rarely – come together while living and working in the same place.

Box 3.1 Gender divisions within a 'local world' of labour

Hanson and Pratt's (1995) fascinating case study of Worcester, Massachusetts shows vividly how labour markets in a particular place are deeply structured by gender relations. The Worcester study suggests there are multiple dimensions to these gender relations that are expressed as distinct sub-local scales of daily activity. Women (especially at the lower end of the occupational ladder) tend to work closer to home than men in order to facilitate time–space regimes that balance both waged-work and various domestic and caring tasks. These geographically constrained employment relations are projected onto an employment map that shows women's (and men's) jobs are concentrated in particular districts. As a result, a household's residential location defines women's access to jobs. This is a distinctly gendered 'friction of distance' leading to the geographical entrapment of some working mothers. Employers also play a role in shaping these gendered employment geographies. Within the workplace, different (gendered) occupational groups may have little contact with one another. At the neighbourhood scale, employers may actively locate their establishments so as to access certain types of employees, a pattern subsequently reinforced by localized, informal recruitment strategies. As Hanson and Pratt conclude (1995: 224), this friction of distance within places constructs multiple highly localized labour markets as 'labour market segmentation is literally mapped onto the ground'.

In summary, as Martin (2001: 461) cogently describes, the labour market of a particular place is:

> a complex assemblage of segmented submarkets, each having its own geographies, its own employment and wage processes ... an assemblage of non-competing submarkets which, nevertheless, are linked together to varying extents via direct and indirect webs of local economic dependency and exchange.

Of course, the range of submarkets will vary significantly from place to place depending on the economic complexity of the localities in question. While a resource industry town in northern British Columbia may only exhibit a handful of segments, a large cosmopolitan city such as Vancouver will be made up of a plethora of more or less connected submarkets. A key determinant will be the nature of the economic activities that are 'grounded' in a particular place, an issue to which we now turn.

PLACING BUSINESS: THE TERRITORIAL NATURE OF PRODUCTION

The localness of business

So far so good: we've seen that workers, along with the trio of groups mentioned in Chapter 2, live essentially local lives. At this point we need to correct a common fallacy. It's frequently thought that while wage-workers may be relatively 'stuck' in place, the businesses they work for are relatively mobile. As Huw Beynon and Ray Hudson (1993: 182) once put it, 'Locations that, for capital, are merely temporary resting places ... become for workers, their families and their friends ... permanent places in which to live'. This view connects with the idea that firms prioritize the 'exchange value' of places over both use and emotional values. Exchange value is the economic return to be had from locating production in a certain place. The idea that firms are generally mobile conjures up the image of them scouring the globe for the most profitable locations. Yet we stated in the first section of this chapter that virtually all types of economic production in capitalist societies are place-based. So how does one reconcile these two seemingly contradictory things? Hudson (2001: 263) provides an answer: 'not all capital is equally mobile, and not all labour is equally immobile'. We have already made mention of the latter regarding migration, on the one side, and those 'trapped' in place on the other. In this section we explain why it's a fallacy to believe that businesses always and everywhere have the capacity to command space – that is, to move between places – leaving workers to their place-based (and sometimes place-bound) existence.

Large firms – notably national and transnational companies (Box 3.2) – undoubtedly have the capacity to go 'place shopping'. The elaborate technical division of labour in these firms, along with the resources (monetary and otherwise) at their disposal, mean that they can search countries, continents and the globe for places that offer the most attractive combination of production, regulation and market conditions. Currently, the top 100 TNCs have 46 per cent of their employees and 42 per cent of their total assets located in foreign countries (foreign, that is, to the 'home' country of those companies) (UNCTAD, 2001). The acclaimed US docu-film *Roger and Me* (1989), by Michael Moore, graphically illustrated this place-switching capacity and its consequences. It depicts the rapid socio-economic decline of Flint, Michigan, after the main local employer – General Motors – decided to relocate car and truck manufacture for the North American market to

Mexico. Companies like General Motors are often acutely aware of place differences in the cost and quality of workers, local infrastructure, relative location, employment laws, state regulations and so on. Even when these companies do not plan to open their own production facilities in new places, they may seek out contractors in these places to supply them with goods and services currently sourced from elsewhere.

However, these facts notwithstanding, even the largest firms do not operate in a frictionless world. Though they may be able to select from among many places when considering where to undertake production, where to source inputs, and where to market commodities, they must still ultimately commit to specific localities. When these companies open production facilities there are enormous 'sunk costs' involved in terms of workplace facilities, investments in the training of workers and so on. Likewise, once new suppliers or markets are sought out in new places, a lot of time, energy, trust and money is invested into making the relationship between firms and supplier/consumers a successful one.

As we will argue below, in much of the current literature on globalization, there has been an undue obsession with the apparently 'placeless' nature of modern capitalist firms, especially TNCs. The obsession is undue not only because these TNCs are only *relatively* footloose – the point we've just made. As importantly, it leads us to ignore the multitude of small and medium sized enterprises (SMEs) that together make up a considerable proportion of the capitalist economy. In most industrialized economies, these firms account for

Box 3.2 TNCs and their significance

While we do not wish to fixate on TNCs as the only firms that influence workers in different places, it is worth pausing to consider their particular significance in the contemporary world economy. Following Dicken (1998: 8), a TNC can be defined as 'a firm that has the power to coordinate and control operations in more than one country, even if it does not own them'. The last clause is important, pointing to the fact that it is not essential for firms to *own* productive capacity in a country in order to be able to determine how that capacity is used. Nike, for example, undertakes much of its production through complex subcontractor networks in East Asia over which it has a great degree of contractual control rather than direct ownership. The scale and significance of TNC activity has increased dramatically over the last 50 years: according to UNCTAD (2001), in 2000 there were some 60,000 TNCs worldwide, owning over 820,000 foreign affiliates with assets of over US$6 trillion. Dicken (1998: 177) suggests there are three important characteristics of TNCs:

- Their ability to coordinate and control various stages of individual production chains within and between different countries
- Their potential ability to take advantage of spatial variations in the distribution of factors of production and in state policies
- Their potential geographical flexibility i.e. an ability to switch resources and activities between places, if necessary at the global scale

In short, TNCs are key shapers of the geography of the global economy through their decisions to invest, or not to invest, in different places. There are two motivations for TNCs to invest in, and switch resources between, different places (Dicken, 1998). While it is useful to distinguish them analytically, in reality many investment decisions reflect the need to balance both sets of factors. First, much investment is *market-oriented*. In other words, investment is intended to serve a specific geographic market by locating within it. Markets may be selected because of their overall size, or because of the particular structures of commodity demand and wealth that exist within them. Secondly, investment may also be *supply or cost-oriented*. Historically, many TNCs in the natural resource industries invested overseas to tap supplies of raw materials such as cotton or oil. More recently, cost-oriented investments have increased in significance. In particular, as Dicken (1998: 189) observes 'many now hold the view that, at least at the global scale, the single most important location-specific factor is labour'. The interplay of a number of characteristics of workers may influence TNC location decisions. Most obviously, there are significant international variations in wage costs, but equally there are geographic variations in the capabilities and productivity of labour, and in the extent of labour 'controllability' due to uneven patterns of collective organization. These attributes play out differently in different places and industries however. Contrary to some apocryphal accounts of globalization (see Preface), not all TNC investment is driven by the need to tap pools of cheap, disorganized, low or semi-skilled manufacturing labour. For example, the majority of FDI in business service sectors such as management consultancy, legal services and advertising circulates within and between high-cost labour markets in advanced economies motivated by a desire to access sophisticated markets and/or pools of specialized skills and knowledge.

over 95 per cent of the total number of businesses, and between 40 and 70 per cent of total employment, depending on the national context (Hayter, 1997). These businesses are often highly localized – think of a corner shop or a chain of regional restaurants. What's more, even though the larger of these business may have a degree of potential geographical mobility (as with the Dyson example mentioned in Chapter 1), they are still relatively immobile compared with the average TNC.

To put all this another way, there's an intractable geographical embeddedness to capitalist production. Firms of *all* stripes must undertake their activities *somewhere*, however many places they can 'shop' among before making final investment decisions. The producing, moving, buying, consuming and servicing of capitalist commodities cannot occur on the head of a pin. Rather, it requires a geographical landscape to give it physical form and shape.

Place, agglomeration and economic growth

Not only does all economic activity have to be located somewhere, but there is also a strong propensity for activities to form localized agglomerations, a phenomenon that has been well studied by economic geographers over the last century. While the emergence of a particular agglomeration is often due to historical accident, once production becomes 'grounded' in a particular place, subsequent growth and development tends to be driven by two kinds of dense localized linkages. First, there may be a broad range of *traded interdependencies* between firms performing different but linked parts of production processes, such as suppliers, partners or customers. By interacting with geographically proximate businesses, firms can minimize the transaction costs involved. Transport costs can be kept low, for instance, in the case of a car manufacturer that is located close to its first tier suppliers. However, co-locating firms need not *necessarily* be functionally related. Hence, secondly, there is growing recognition that firms within an agglomeration may also benefit from a broad range of *untraded interdependencies* (Storper, 1995). Some of these less tangible advantages are economic in nature, such as the emergence of a suitable common pool of labour, while others have a significant socio-cultural component. For example, there may be a range of place-specific conventions, informal rules and habits that help coordinate economic action among contiguous firms. Investment bankers in the City of London, for instance, may gain important forms of business knowledge through informal networks that operate as much through bars, clubs and restaurants as formal workplaces. Localities, then, act as places of sociability and face-to-face contact, and these interactions are crucial in enabling the knowledge transfers and innovation upon which economic growth depends.

There are, of course, many forms that the agglomerations of economic activity may take. One such form that has received much attention from social scientists since the 1980s is the so-called 'new industrial district' based upon localized cooperative networks of innovative small firms. Also termed 'Marshallian' districts after the pioneering work of economist Alfred Marshall on the topic in the early 20th century, the concept emerged from empirical analysis of the successful expansion of craft industries in the small towns of Emilia-Romagna, Italy. However, there is now recognition that the Marshallian district is but one form of dynamic agglomeration. Economic geographer Ann Markusen (1996), for example, identifies three further forms of industrial district in her essay on what she evocatively terms 'sticky places' (see Figure 3.3). First, she notes the occurrence of 'hub-and-spoke' districts, whereby a few major local firms in a small number of industries heavily influence the local industrial structure. These vertically integrated firms are surrounded by a local web of smaller and less powerful suppliers, as in the case of Boeing and Microsoft in Seattle. A second type of district is the 'satellite platform', essentially a cluster of branch plants of externally owned multi-plant corporations, as in the case of Singapore's electronics industry. A third type is the 'state-anchored' district, whereby a key public entity (university, research laboratory, or defence establishment for example) is the key anchor tenant. Examples here would be cities in the United States that have state universities and/or are a state capital such as Madison, Wisconsin and Ann Arbor, Michigan.

While such typologies hint at the potential variability of agglomerations, in reality of course the nature of production is highly complex in most places. Two final points can be made in this regard. First, in many places several different types of industrial district coexist and overlap, creating 'sticky mixes'. In other words, 'real world' places are amalgams of different forms of agglomeration. Elements of all four of Markusen's types of industrial district can be discerned within contemporary Silicon Valley, for example. Thus while some territorial production complexes will take their character from just one or two industries, many places are a complex assemblage of several different formations. Second, we need to reiterate an earlier point: namely, that places are ever changing and dynamic – due in part to the capitalist imperatives of competition and innovation – and that their very character will vary over time as the various territorialized linkages they contain wax and wane.

LOCAL SOCIAL REGULATION

What we've seen thus far in this chapter is that workers' daily existence is necessarily local, while capitalist firms can rarely escape the

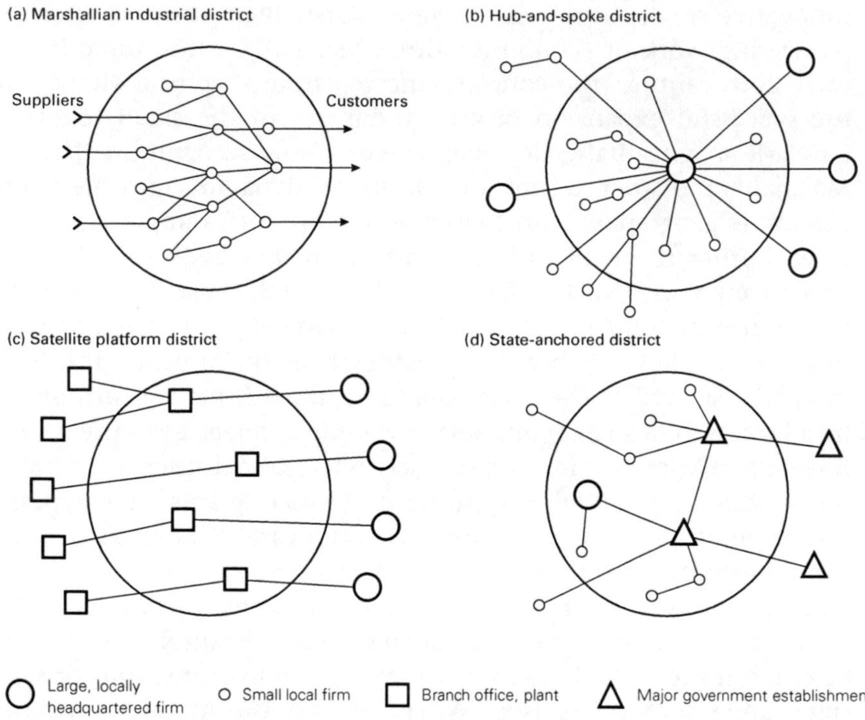

Figure 3.3 Key types of industrial district (adapted from Markusen, 1996: figure 1)

pull of place – even relatively mobile ones. But places are about far more than just workers and firms, though the two are clearly important. As we saw in Chapter 2, the employer–employee relationship is socially regulated. So to understand the importance of place we also have to comprehend why social regulation is always expressed in place and why many regulatory institutions are as place-based as workers and firms are. Thus place can be thought of as 'the sphere where local and non-local (national and international) systems of rules, norms, customs, legal structures, and regulatory mechanisms intersect to shape and institutionalize the behaviour of both workers and employers' (Martin, 2001: 464).

In essence, the reason why the social regulation of workers and firms is necessarily local is because of what we have discovered already in this chapter: that is, that workers and firms are, at some level, *always* place-based. There are two important dimensions of the 'local' nature of regulatory processes. On the one hand, many institutions and therefore mechanisms of labour regulation are locally exclusive: they are only local in scope. There are several aspects to this (Peck, 1996).

Local state actions obviously play an important role through a range of policies related to child care, economic development, immigration, and the delivery of training, education and social services (Cockburn, 1977; Cochrane, 1993). Local household and family structures will shape gender divisions in waged, caring and domestic labour. Local agreements between unions and businesses will affect the nature of collective bargaining. The local workforce will have distinctive characteristics in terms of aspirations, turnover patterns and traditions of solidarity. Finally, the local institutions of civil society will mobilize around place-specific issues and problems. For example, various community groups may team up to contest the environmental impacts of a local factory. Taken together, we see that a broad range of regulatory interventions will be initiated and implemented locally. On the other hand, national and international regulatory processes will adopt particular local forms due to their interplay with the institutional and economic specificities of any given locality. These specificities may include, for example, labour force characteristics such as segmentation, consequences of previous political action, the nature of the local built environment, and the relative strength and vitality of local state and civil society institutions. Regulatory mechanisms concerned with labour legislation, training and job creation programs, welfare systems and taxation structures, for example, will interact differentially with these distinctive local forms. For example, the costs of implementing national minimum wage legislation may be far greater for employers in a struggling local labour market characterized by high rates of unemployment and low wages, than in a very tight labour market with low unemployment and already relatively high rates of pay.

For Peck (1996: 107), these two dimensions together create what he calls 'a distinctive regulatory milieu' in localities. To be clear, recognizing that the social regulation of labour varies between places is not the same as saying that the *control* of labour regulation resides at the local scale. We will return to this significant issue in the next chapter. For now, the important point is that the regulation of wage-labour is both *locally expressed* and *locally variable*.

GROUNDING THE WORKER–EMPLOYER NEXUS

Local social relationships

If we combine the insights of the previous four sections, we begin to see that the employer–employee relationship in capitalist societies has a manifestly local dimension to it. What's more, we've seen that it's too simplistic to assume that all workers are place-bound, while businesses are merely place-based. What we can say, though, is that at some point virtually all workers and employers *must* develop specifically local relationships with one another. They have to engage their various power repertoires *in situ* to try and iron out differences in needs and expectations.

However, it's not simply that employment relations occur 'in place'. Rather, it's that the placing of these relations *materially affects* how they unfold. This localness can determine how the various competing power repertoires play out, and thereby fundamentally affect the terms and conditions under which labour power is used and reproduced. For example, employers will often adjust their recruitment strategies, working practices and wage rates to reflect local conditions. This may be either favourable or unfavourable for the concerns of workers. In places where labour markets are tight, employers may alter recruitment practices to ensure the continuation of production. For example, in certain labour intensive industries such as clothing, employers may promote homeworking to tap new pools of female labour. For workers, tight labour markets may offer great opportunities to regularly switch jobs and move up the employment ladder quickly, as evidenced in the burgeoning high-tech labour markets of Silicon Valley in the early 1990s. On the other hand, in places where labour supply exceeds demand, employers may take the opportunity to restructure, by shedding jobs and investing in new equipment for example. This is often the situation when new employers enter heavily deindustrialized localities. Hudson (2001: 122) describes how Nissan received some 33,000 applications for 600 jobs at their Sunderland car plant in the late 1980s. In such situations, employers can be very selective about who they choose to employ and, indeed, about the ongoing terms and conditions of employment. The power repertoires available to workers in these circumstances may be few. Clearly, place becomes a crucial 'container' for employer–employee dynamics: depending on the context, it can either enable or constrain the two parties.

What's more there's local *variability* in the character of the employer–employee relationship. Peck (1996) explains the uniqueness of labour markets in different places through the intersection of two dimensions. On the one hand, labour market dynamics are locally constituted through unique intersections of labour demand and labour supply: the production–reproduction dimension. Put simply, places vary dramatically in terms of the number and type of employers they contain, and equally widely in terms of the number and range of workers that are present. On the other hand, as we have already seen, labour market institutions vary geographically in terms of their nature, functions, and the local impacts of their strategies: the regulatory dimension. We need to be careful here however. The argument is not that labour market processes are *singular* – that is, completely unrepeatable from place to place – but rather *unique*: that is, 'in their operation and empirical effects, they work in different ways in different local contexts' (Peck, 1996: 266). A new manufacturing technology, for example, may place certain types of worker under pressure across a wide range of places. How exactly the pressure is expressed and reacted to, however, will be *place-specific*. What we have is a series of unique contexts within which the strategies of employer and employees are formulated and regulated.

The geography of class identity and class action

The consequence of all of this is double-edged for workers. On the one side, place can undermine – or at the very least complicate – the wider class consciousness of workers. In a world of multiple workers and firms in multiple places, classes in-themselves are geographically fragmented. In Hudson and Sadler's (1986: 189) words, 'a concern with more general class solidarity, even if this is recognized, is [often] subordinated to a more immediate concern with living and working in a ... particular place'. Accordingly, the possibility that they become classes for-themselves may be challenged by the local nature of any emergent class consciousness. Workers and their organizations are always, in the first instance compelled to respond to specific *local* employment, wage or unemployment problems. Such campaigns to 'defend places' can vary greatly in their form, as we'll see in Chapter 6, from politically progressive protests that unite workers in different places to those that pit place against place in a regressive struggle. In the latter case, the threat of losing jobs and investment to other places can become the catalyst for a campaign that brings together the interests of certain local capitalists and managers with local workers, local politicians and local institutions of civil society. Capitalists or their representatives may be compelled to join these projects in order to preserve the conditions necessary for producing profits in that particular place. Whatever the local outcome of such a 'cross-class alliance', it nonetheless contributes to intra-class geographic fragmentation among workers in the wider national, international and global economy.

On the other hand, however, the local nature of worker existence and identity is not *simply* a hindrance to the formation of a fully-fledged class consciousness among workers worldwide. The classic Marxian view is that once different workers come to realize they have a common interest in challenging the rule of employers, then they will be able to achieve emancipation by uniting in action. However, the problem with this view is that it assumes that place differences in worker identity must be overcome, as if they're less 'authentic' than a wider inter-place class identity that's somehow waiting to burst forth. Yet, surely, place is what *makes* classes flesh, both in- and for-themselves. It's not as if class is a non-geographical phenomenon that simply happens to be expressed in a place variable way. Rather, it's precisely this variability that *constitutes* class in the real world. In Hudson's (*op. cit.* 223) words, class 'is geographical in the first rather than last instance'. In consequence, it's a mistake to think that workers need to throw-off the 'shackles' of place if their 'real' social interests are to be realized. Commitments in and to place by workers, be they with or against their employers, are part and parcel of what everyday life in a capitalist world is all about. This has some important implications. We argued in the last chapter that if workers are to develop a wider common awareness of their class position and class interests, a major *effort* of consciousness building is required. This effort, it should now be clear, needs to be as much geographical as social. It involves negotiating not just social differences of an intra- and extra-class nature, but also place differences. In other words, it involves the construction of what Richard Hyman (1999: 94) calls 'imagined solidarities', where working people 'perceive a commonality with others who they do not know and of whose specific identities they are unaware'.

SUMMARY AND PROSPECT

In this chapter we have shown how workers, firms and regulatory institutions are all tied to place in many and complex ways. As a result, the nature of employment relations and their regulation is intrinsically both local and locally variable. In turn, the various power repertoires mobilized by employers and workers are locally variable and locally articulated. In this sense, the exercise of worker (and indeed employer) agency is *always* contingent on local circumstances. In some instances local relations can hinder effective action, and in some cases can strengthen it. Place, then, matters profoundly. However, the relational approach to place that we have advocated in this chapter also recognizes the place interconnections and interdependencies that help to shape 'localness'. Hence we need to develop our argument further in the following chapter by considering relations between places: *across* space, and at different geographic scales. For, as we argued at the start of Chapter 1, places are no longer insulated from extra-local forces.

four

Spacing and Scaling Labour

In Chapter 3 we saw how workers, production, and regulatory mechanisms are all very much 'placed', in the sense that they ultimately interact locally and in the workplace. We now move on to develop our geographical argument in four stages. First, we show how translocal relations and institutions stretched over a wider space are increasingly influencing what happens at the local level. Second, we argue that translocal relations and institutions are scaled from the local to the global, and that this scaling decisively affects what happens to workers locally and what workers can do about it. In effect, geographical scale specifies the content of any given place–space relationship. Third, we use the concept of the 'local labour control regime' to integrate these 'spacing' and 'scaling' arguments with those made about the 'localness' of labour in Chapter 3. Finally, we offer some preliminary comments on the geography of worker politics, arguments that will be developed substantially in the rest of the book.

FROM PLACE TO SPACE (AND BACK AGAIN)

Ours is a world of 'rooted translocalism' (Katz, 2001: 724). We've argued that wage-workers, their families and friends, employers, retirees, the unemployed, firms and regulatory institutions combine in different constellations in myriad localities. We've further argued that inter-place relations produce not simply geographical differences but real unevenness in the circumstances of everyday local life. The cultural critic John Berger (1974: 40) once famously said that 'it is space ... that [now] hides consequences from us'. What he meant is that the stretched out relations and translocal organizations connecting places are usually invisible to us, even though their effects are quite real. Actions and events in any given place are made possible by what sociologists call 'unacknowledged conditions' and they can have 'unintended consequences' elsewhere. In Anthony Giddens' (1990: 19) words, '... the visible form of the locale conceals the distanciated relations which determine its nature'. So what are the principal types of space relations and organizations that we need to make transparent if we're to understand what's happening to wage-workers in place – whether or not these workers are aware of them? Having spent the last chapter emphasizing what makes socio-economic life persistently local, and having sketched a general argument for why places cannot be insulated from others in theory or practice, it's now time to say more about the translocal ties that help make contemporary places what they are. We can distinguish translocal ties relating to workers, firms and regulatory institutions respectively. In each case, these ties can either be competitive or cooperative, intentional or unintentional, conscious or unconscious. Though we separate them out here analytically, in reality they combine in complex and changeable ways. Together they form what Cox (1998) calls 'spaces of engagement' that to different degrees resonate with the 'locally dependent' actors, events and institutions mentioned in the previous chapter.

Worker inter-connections and inter-dependencies

There are four principal categories of worker inter-relations and dependencies to consider:

- **Competitive production relations** – because of the local nature of work, reproduction and identity, and because capitalism is predicated on economic competition, this generates geographical competition. As Hudson and Sadler (1986: 179, emphasis added) once put it, 'Seen from the point of view of workers, their families and dependents in such places, the only feasible solution often appears to be to fight for *their* factory, *their* mine, for *their* community or region ...' Workers in different places are thus competing with distant strangers for jobs, investments and markets, despite their erstwhile similarities as members of a wider labouring class, wider class fractions or wider occupational groups. This competition can occur between established centres of commodity production or between old and new places of economic activity.

- **Cooperative production relations** – in contrast to the previous point, because the cost of living, the average wage, the living wage and so on are place-variable, workers often have an incentive to cooperate with others elsewhere in order to secure a fair deal for *all* from employers and regulatory institutions. As Richard Hyman (1999: 99) notes, 'Without differentiation there would be no need for solidarity.' These relations of geographical cooperation have historically been formalized through national and international trades unions (see the 'Translocal regulation' sub-section below). Using the combined power of local union memberships, these translocal unions can push for inter-place equity in worker pay, conditions and rights. In other cases, workers can use non-union organizations to pursue issues translocally.
- **Migration relations** – the migration of people in search of wage-labour is, numerically, at an all time high in the early twenty-first century. Intra- and international migrants criss-cross social space, impelled by a mixture of job unavailability in their home places and labour demand in their destination areas. When migrants send remittances back to their areas of origin, the paid employment in one place is, effectively, financing social reproduction many kilometres away. But the geography of migration is not a well-oiled machine. For example, forced migrants often end up in places of labour over-supply, while places of high labour demand are frequently unable to attract the necessary influx of outside workers. In addition, migrations relations can lead to competition among workers, depending on the circumstances. For instance, if migrants are willing to work for lower wages than resident workers, then this can drive down the salaries of the latter or even cost them their jobs.
- **Consumption relations** – commodity movement and consumption is at once stretched out and local. Though workers in one place do not buy commodities in order to help keep distant workers in their jobs, this is an unintended outcome of locally purchasing and using commodities that hail from afar (Box 4.1). Likewise, if those working in one place purchase commodities made elsewhere under harsh or oppressive conditions, then they inadvertently help perpetuate those conditions.

Box 4.1 The hidden geography of consumption

Wage-workers are commodity consumers as much as commodity producers. Yet in buying and using commodities workers rarely realize that they are thereby helping to sustain the livelihoods of workers elsewhere. This is hardly surprising. When one handles a commodity in a shop, say, it does not tell us who made it and under what conditions. Though clues are given by labels and packaging they rarely offer real insights into the working lives of commodity producers. Consider, for example, a commodity that is considered to be quintessentially English: tea. Tea remains among the most popular household drinks in England. Yet, in terms of its physical production, is it scarcely English at all. If we take the Twinings tea bags served in our university canteen, a close inspection reveals a hidden, international geography of production that is obscured when we drink tea with our lunch. To begin, Twinings tea might begin life half a world away on tea plantations in Sri Lanka, Kenya and India. There workers are paid perhaps 90 pence per day to pick 19 kilos of tea, a mere fraction of which will go into a 50 tea bag box of tea sold in England for £1.70. Once dried, rolled, fermented and baked, the tea is sent to Twining's factory in Andover, in the south of England. There it is put into tea bags made from pulp derived from Scandinavian and American tree plantations and polypropylene from Fina Oil's Belgian chemical factory. Before appearing on supermarket shelves and in university canteens, the tea is put in boxes specially made in a Gloucester factory which are printed with the Twinings logo in Maidenhead prior to being wrapped and sealed in Twinings's assembly plant in north Tyneside. What all this means is that each time an British person buys Twinings tea they are, unbeknownst to them, entering into a set of complex, geographically stretched relations with multiple workers and helping to sustain their jobs. Equally, by not buying other brands of tea they may be inadvertently jeopardizing the livelihoods of workers in other tea companies (*source*: *The Guardian*, 25 June 2002).

Firm inter-connections and inter-dependencies

Due to their ability to straddle, and in some cases switch between, different places, the inter-place connections among firms are arguably more numerous and complex than those identified in the case of workers. Here we delimit six key types, the first four of which refer to inter-firm network relations, the last two to intra-firm networks:

- **Inter-place supply relations** – while firms often source many of the inputs they require locally, they will nonetheless also rely on a range of inputs generated in other places. These may be material inputs, for example in the form of individual or partly assembled components, or they may be non-material inputs to the production process such as financial or consultancy advice. These inputs may be secured through 'arms-length' transactions on the open market, or through more tailored subcontracting relations.

- **Inter-place market relations** – some small and medium-sized firms may survive by selling their products or services into an immediately 'local' market. Most firms of a reasonable size, however, depend on accessing markets in other places. Depending on the type of activity in question, this may take the form of an intermediate market made up of other businesses (thereby intersecting with the supply relations described above) or a final market where commodities are offered to individual consumers (thereby intersecting with the consumption relations of workers described above).
- **Inter-place cooperative production relations** – aside from the 'vertical' networks described above, firms may also enter into 'horizontal' networks with partners in other places. While many progressive forms of collaboration are known to take place locally (as noted earlier), others take place over greater distances up to, and including, the international scale. These strategic alliances occur for three main reasons: to pursue collaborative research whilst spreading the associated costs and risks; to share and occasionally produce complementary technologies; and to expand markets through joint distribution, licensing or marketing arrangements.
- **Inter-place competitive production relations** – another form of 'horizontal' network relation is that between competitors. As with partners, competitors may be local, but equally they may be based extra-locally. Competition is usually meant to denote a contest for market share based upon parameters such as price, quality or brand image, although equally competition may occur for spatially mobile inputs such as skilled workers or venture capital. Competition often takes the form of one firm taking over another (a so-called merger and acquisition). Somewhat counter-intuitively, these competitive relations can occur simultaneously with the cooperative relations described above, as evidenced in research collaborations between leading computer manufacturers.
- **Intra-firm inter-place relations** – while some multi-locational firms will have various plants within a single place, most will have branches or subsidiaries in a range of different places. Thus there are a variety of intra-firm networks that bind and connect workplaces. These may take a variety of forms: firms may be simply replicating the same activity in a variety of sites; some sites may be generating inputs for other workplaces; sites may be engaged on a common project, or sites may be responsible for entirely different activities within the firm. Hence, depending on the context, the nature of the

interdependency may range from simply working towards the profit goals of a common corporate entity, to intense exchanges of knowledge and capital through working on a joint project.

- **Intra-firm inter-place 'switching'** – finally, we need to reiterate that some firms have the ability to potentially alter the geography of their operations either as part of a planned strategy, or in response to some form of external threat or crisis. To simplify what in reality are extremely complex processes, firms may either chose to make *in situ* adjustments by expanding, contracting or altering the nature of operations in their existing network of workplaces, or engage in *ex situ* locational shifts through closing a plant, opening a plant, acquiring a plant or – in extreme cases – moving an entire plant. In reality, while millions of firms are exposed to the exigencies of global inter-place competition, a much smaller proportion respond by undertaking *ex situ* restructuring. In many cases, locational shifts are complex and require long-term planning and massive investment. For many single-site small and medium sized firms, moving their activities is simply not an option.

Our understanding of these firm interconnections can be advanced through Massey's (1984) already mentioned concept of spatial divisions of labour. In a general sense, the term simply reflects that different places are characterized by different clusters of firms and workers. Here, though, we will explore how these spatial variations are both *created* and *exploited* by the ongoing restructuring of capitalist firms in their pursuit of profit. Hence, through varying combinations of the six forms of interconnection described above, production is organized systematically *across* space.

Massey's analysis is complex and detailed, but following Peck (1996: 156), we can summarize her analysis in five key stages:

- Growing global competition is driving corporate restructuring.
- Corporate restructuring is leading to the separation of the 'control' functions performed by managerial workers from the 'execution' functions undertaken by manual workers, as pre-existing labour processes and production systems are 'broken up'.
- This functional separation of 'control' from 'execution' increasingly is an explicitly geographical process, with the different functions being located in places that correspond with the required labour supply. Most typically, low-skilled manual tasks are located in peripheral areas, while managerial and R&D tasks are undertaken in core cities and their immediate hinterlands.
- Over time, as different processes of corporate restructuring ensue, 'layers' of investment spread across the economic landscape, creating new

production geographies and redefining the nature of relationships between places as they gain or lose different kinds of economic activities. In particular, shifting patterns of 'control' functions will alter relations of dependency between different places.

- New spatial divisions of labour are created through the complex interactions between a new layer of investment – say in software development centres – and the pre-existing economic make-up of a particular place. These interactions, or 'combinations' as Massey calls them, will take on different forms in different places.

Hence, spatial divisions of labour is a concept that reveals the reciprocal relations between trans-local processes of economic restructuring and the particularities of places. This framework can be mobilized in two important ways. First, by focusing on a particular firm or sector of economic activity, we can understand how production is organized and re-organized spatially in line with changing labour requirements and competitive conditions. Second, the economic fortunes of particular places can be thought of in terms of the position of that locality within successive spatial divisions of labour (see Box 4.2 for a consideration of the UK case).

Box 4.2 Spatial divisions of labour in the UK economy

If all this sounds a little abstract, Massey's theory was grounded in a detailed empirical analysis of spatial restructuring in the UK economy in the 1960s and 1970s. She charted the emergence of a new spatial division of labour in manufacturing based upon a separation of skilled and managerial tasks (performed in many cases in London and its environs) from unskilled manufacturing work done in so-called 'branch plants' (located in areas such as Northeast England and Southeast Wales). The choice of these peripheral regions as locations for branch plants was far from accidental: rather, it represented a desire on the part of employers to tap into large pools of relatively cheap, disorganized, and in many cases female labour to perform routine assembly tasks. Such labour was available in sufficient quantities in regions being decimated by dramatic declines in male-dominated 'traditional' industries such as coal and steel production. In this way, the new spatial division of labour was interacting with a previous layer of investment characterized by 'regional specialization', with towns and cities being associated with particular sectors, for example finance in London or textiles in Manchester. The analysis can be brought right up-to-date by considering how peripheral regions of the UK have had to move towards various forms of service sector growth – such as call centres and urban/heritage tourism – as branch plant manufacturing has declined since the early 1980s in a context of increased global competition.

Figure 4.1 illustrates in a simplified way the 'cloning' and 'part-process' corporate structures that lie behind different spatial divisions of labour. In the cloning structure, the production apparatus is simply replicated in different localities, with ultimate control residing in a HQ located in one particular production branch (usually the initial place of origin of the firm). In the part-process format, there is a technical division of labour between branch plants, with components being made in one location and passed on to another for final assembly. Three points are important here. First, the different stages of the production process may have different labour requirements, and hence tend towards different kinds of location. Second, as alluded too earlier, these spatial divisions of labour can be constructed through either inter-firm (externalized) or intra-firm (internalized) networks. Third, and most importantly, this part-process format increases the level of interdependency within the system, as assembly in one place cannot occur without the supply of components from another. Moreover, it reveals that inter-place relations of *control* are extremely significant in shaping worker opportunities and livelihoods, and need to be added to our understanding of inter-place *cooperative* and *competitive* interdependencies. What is particularly crucial here is that vital decisions about the ongoing operations in a particular place may be being taken in another locality.

Translocal regulation

Though we've already seen that some regulatory institutions and mechanisms are either local or else find local expression, many are also translocal. We can distinguish translocal worker institutions, translocal firm institutions and translocal non-worker/non-employer institutions (like the state) that influence the employee–employer relationship at the local level. We have already flagged the importance of the state as a key translocal regulator of labour markets. On the worker side, most unions are part of a national organization, and are increasingly networked together internationally. Equally, firm institutions such as trade associations are usually organized at the national scale, and 'local' organizations such as Chambers of Commerce are typically members of a national association. The nature of labour market regulation in a particular place will ultimately depend on a complex interplay of both local and extra-local regulatory bodies and mechanisms. In short, there are two key dimensions to translocal regulation:

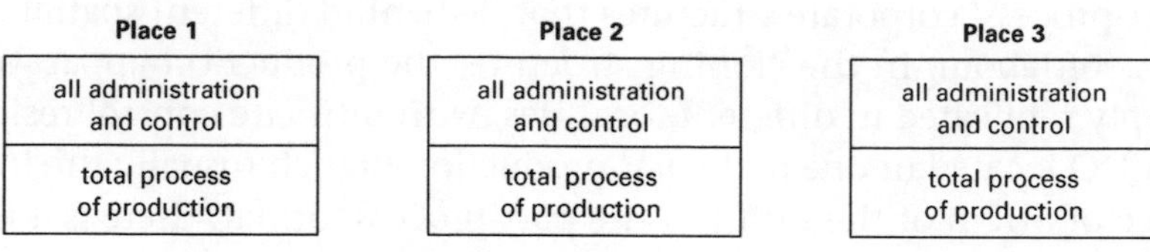

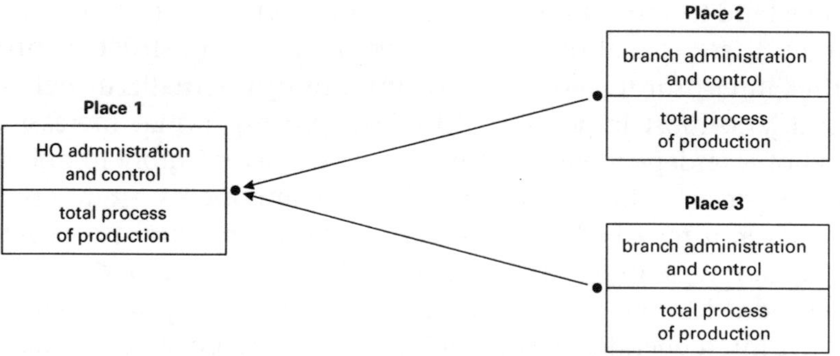

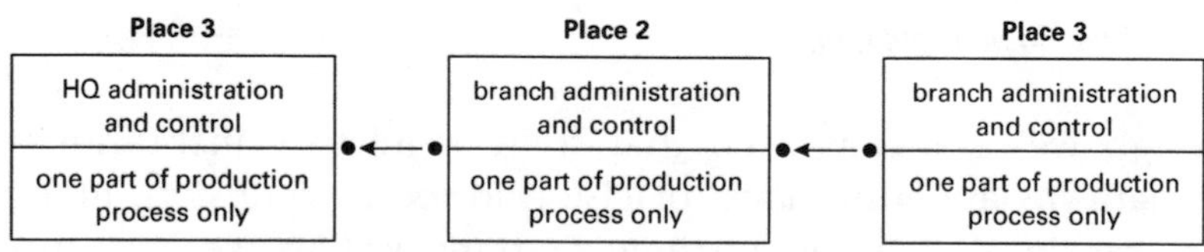

Figure 4.1 Three spatial structures of production

Source: Massey, 1995c: figure 3.3

- **Extra-local regulatory institutions** – first, many local institutional bodies are simultaneously geographically distinctive *and* embedded in national and international institutional forms (Peck and Tickell, 1995). This relationship may be fairly loose, in terms of membership of a national institution, or tighter in the case of local branches of a wider translocal body. An analytical emphasis on the local should not obscure the influence of regulatory control imposed at the national (and sometimes international) scale that can impose structural constraints upon, and shape the very nature of, local regulation. For example, the freedom of the local arm of a government training agency to develop place-specific programmes will ultimately be constrained to some degree by the objectives of the national organization of which it is part.

- **Extra-local regulatory mechanisms** – second, and following our earlier argument, many regulatory mechanisms are implemented translocally. The state may, for example, implement *national* health and safety guidelines, or a *national* minimum wage policy. Groups of states, through quasi-state bodies, may seek to do the same at the international level. The fact that the impacts and outcomes of these policies will be different in different places does not alter the fact that these are regulatory measures designed and implemented by extra-local regulatory bodies. Some of these 'translocal' measures may – either implicitly or explicitly – be targeted at certain kinds of places, for example policies aimed at job creation in areas of high long-term unemployment, or a 'national' sectoral policy that only really impacts in areas where the industry in question is located.

When we add the above three sub-sections together, we begin to see how a dynamic process of *inter-place competition, cooperation and control,* and *movement of firms and/or workers* is endemic to the contemporary capitalist world. As we've outlined, geographical competition can take two main forms. The first is of existing places (or rather place-based actors) acting reactively or proactively to compete for jobs, investment and markets with other existing places. The second is when 'new' places arise in areas that have hitherto had little experience of wage-work or commodity production and compete for national, international and global markets. In the case of geographical cooperation, both workers and firms can derive some real benefits from strategic short- or long-term alliances with distant actors. Finally, in terms of capital/labour mobility, firms and workers can sometimes 'escape' the confines of one place in order to relocate in another that offers more favourable conditions. In each instance, the impulse to compete, cooperate or to move is fundamentally attributable to the way capitalism establishes the ground rules for daily action within and between places. Though places are not *reducible* to the logic of capitalism (there's much more to them than this), that logic nonetheless frames the place-based and inter-place activities of different workers, firms and regulatory institutions.

The moral of the story is, it seems to us, this: if workers are to defend and to enhance their livelihoods, they must try to harness, rather than shut out, those spatial relations that are in part constitutive of their home places (though towards the end of Chapter 6 we explore some exceptions to this). In Ash Amin's (1999: 42) words, '...finding a place in a world of "stretched and deepened" time-space connectivity might now be more a matter of how connectivity is negotiated or worked to your advantage, rather than resisted ...'. Where such negotiations are undertaken through purely local responses and fail, then workers must, in a sense, exert themselves over space. They can do this by actively uniting with workers elsewhere to undertake coordinated translocal actions that benefit some or all involved. Alternatively, they can move across space in order to find new places of employment. In each of these cases, workers must try to use what powers they have at their disposal, or else they are no more than passive players in the making of the modern capitalist landscape. This exercise of worker agency can take myriad forms. At times it can involve locking horns with local or non-local employers and regulatory institutions. On other occasions, it can involve cross-class alliances and institutional cooperation. But in almost all cases it involves a mixture of action at the local and non-local scales, rather than one scale or the other (a point we'll return to in the next section).

THE DIFFERENCE THAT SCALE MAKES

In the previous section we've categorized some of the translocal relations and institutions that connect otherwise different place-based workers and communities. We want, in the rest of the chapter, to explain why and with what consequences these relations and institutions are organized at several *geographical scales*. Scale is the key to understanding what, in reality, is done to wage-workers in places and what these workers can do to help themselves. This is the third term in our geographical triad. It helps us to understand how the space–place relation unfolds in any given situation. Though we've touched upon it here and there in this chapter, it's essential that we now explain the nature and consequences of geographical scale. In this section we start with some general reflections, before classifying the various scales at which contemporary socio-economic life is organized.

There are at least two common misapprehensions about the nature and importance of geographical scale. The first is that scale is given and natural. Thus when we use terms like the 'local scale' and the 'global scale' it seems as if we're discussing things that are absolute and unchanging. However, geographers like Neil Smith (1993) and Erik Swyngedouw (1997) have argued that while *cartographic scale* may indeed be a fixed metric (as on a map), geographical scale is anything but this. For Smith and Swyngedouw geographical scales are *socially constructed*. In other words, they are the product of social relations, actions and institutions. Thus the 'national scale', though a seeming given, is in fact a relatively modern phenomenon. It only came into being as a meaningful and widespread scale at which certain social relations and institutions were organized from the late eighteenth century. What's more, the content of the national scale – that is, the specific mixture of things that give rise to and are 'contained' within it – is highly variable between countries (compare the United States, say, to Nigeria). This brings us to the second misapprehension about scale: namely, that it's of merely technical interest. In this view, regardless of whether scale is socially constructed, it is seen as being of little material consequence. So the fact that certain aspects of social life are organized at certain scales, but not others, is regarded here as being a matter of detail rather than importance. Against this, theorists like Smith and Swyngedouw have insisted that geographical scales do not merely 'express' social relations but *actively enable and constrain them*.

This idea that geographical scale is of *causal* importance has already been touched upon in this chapter. For instance, we've seen that the place-bound nature of certain workers – that is, their inability to organize or move over space – can make them vulnerable to the actions of certain firms. But scale does not just confine, it also facilitates. The power to do particular things at specific scales can be vital to achieving an individual's, group's or organization's objectives. This is not necessarily to imply that the 'bigger' the geographical scale at which social actors can organize themselves the more power they have at their disposal. As Swyngedouw (1997: 141, emphasis added) rightly observes, 'the ... priority never resides at a *particular* geographical scale ...'. Thus, while a TNC *may* have power over certain workers because of its geographical mobility within, say, a continent, workers are not *always* relatively less powerful because their lives and actions are usually place-based. Sometimes, place-based actions in one locality can allow a given group of workers to realize their objectives, just as the fact that certain regulatory institutions are globally organized does not mean that their regulatory mechanisms will be equally effective in all the places where they are expressed. What this means, then, is that the importance of acting at certain scales is always *relative and contingent,* not structured hierarchically from the local to the global.

This brings us to a final general comment about geographical scale. Certain relations, actors and institutions in our capitalist world are multi-scalar, while others are highly scale specific. For instance, we've already discussed the 'locality dependence' of things like worker reproduction. But the scale at which social life is organized is not unchanging. Entirely new social entities can be forged at certain scales. A good example is the World Trade Organization, which regulates the global traffic in commodities between places and countries. The WTO is only a few years old, and is the most formal expression of various postwar attempts to manage world trade by applying rules that all trading nations must adhere to. Alternatively, when actors, relations or institutions that are typically contained at one scale find temporary or permanent expression at another, we can talk of 'scale-switching'. Good examples of this re-scaling process are the recent attempts by many workers to construct both local and international and global alliances to put pressure on their employers (we consider these in detail in Chapters 6 and 8).

Thus far we've used the language of place (the local scale) and space (the translocal scale) in this chapter, with the brief mention of the sub-local scale. It's now time to offer a more systematic survey of the scaling of social relations and institutions in the contemporary capitalist world, a scaling which has arguably been profoundly reworked, in Western countries at least, over the last 20 to 30 years. We consider the scaling of firms, regulatory institutions and workers in turn. There are rapidly growing literatures on each of these three topics, so all we can do here is sketch out the key dynamics, many of which will be returned to in more detail in subsequent chapters. Each section is accompanied by a figure (Figures 4.2, 4.5 and 4.6) that represents the key scalar structures in question, and indicates how they are being 're-scaled' in contemporary capitalism. In all cases the most basic scalar unit is the workplace, for, as we have already described, this is ultimately where the relations between employers, workers and regulators play out on a day-to-day basis.

Re/scaling firms and the economy

The globalization literature, as we have already argued, is far from perfect. It does, however, draw attention to the key contemporary re-scaling of firm activity, namely that which is occurring as increasing numbers of businesses coordinate their production transnationally. While in

1950 the world economy could be described as a series of relatively self-contained national economies engaged in 'arms-length' trade, a far more interconnected and integrated global system has now developed. This activity is reflected in growing levels of international trade, foreign direct investment and capital flow, and the growing number of TNCs, international strategic alliances and international subcontracting networks (Figure 4.2). The following statistics give an impression of the magnitude of this 'upscaling' of economic activity, a process that has accelerated significantly during the last 20 years:

- World trade expanded more than threefold over the period 1980–2000, with exports of merchandise and commercial services rising from US$2400 billion to US$7882 billion (www.wto.org).
- Global stocks of foreign direct investment grew tenfold over the period 1980–2000, rising from US$616 bn to US$6314 bn (UNCTAD, 2001).
- Cross-border mergers and acquisitions have grown substantially: for the year 1987 the total value of such transactions was US$74.5 bn, but in 2000 the figure was US$1143 bn (UNCTAD, 2001).
- International strategic alliances now proliferate: just over 42,000 cross-border deals were signed in the period 1989–99, accounting for 68 per cent of all strategic alliances (Kang and Sakai, 2000).
- The number of people employed in the foreign affiliates of TNCs increased from around 17.5 million in 1982 to 45.6 million in 2000 (UNCTAD, 2001).
- The daily turnover of foreign exchange on the world's currency markets was US$1,500 billion in 1998, as compared to US$590 billion in 1989 (BIS, 2001).

The new global economy is also characterized by the growing geographic complexity of these transnational flows and linkages, with more and more – but by no means all – countries and sectors being brought into their purview (see Dicken, 2003, for a detailed geographical analysis).

These dynamics reflect profound changes in the nature of corporate spatial divisions of labour, a concept introduced earlier in the chapter. These divisions of labour can be organized at all *spatial scales*. Many small and medium sized businesses, for example, operate from one site, or perhaps several within a locality. As with the dynamics that Massey described in her book, many multi-plant firms still operate through intra-national divisions of labour. However, one of the defining characteristics of the global economy over the last two to

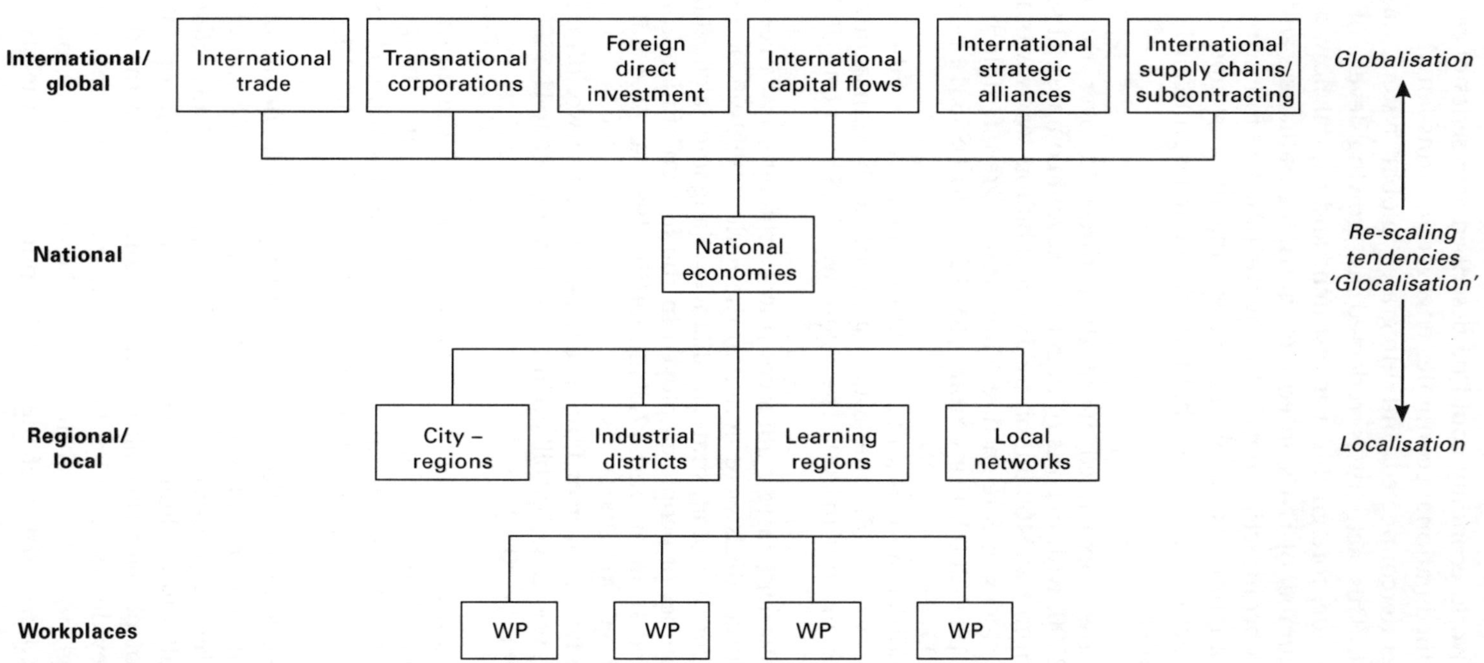

Figure 4.2 The re-scaling of the economy (adapted from Swyngedouw, 2000a, figure 27.3)

three decades has been the dramatic increase in the number of firms (TNCs) organizing their spatial divisions of labour at the *international* scale (see Box 4.3 for a review of attempts to conceptualize these dynamics). Four further characteristics of these international divisions of labour need noting:

Box 4.3 IDL, NIDL or GDLP?

Here we will briefly review three attempts to conceptualize international spatial divisions of labour, which coincide broadly with different phases of development of the global economy.

- First, the theories of classical economists, and in particular Ricardo's notion of comparative advantage – which stated that countries would specialize in producing goods for which they had a comparative cost advantage, and import those for which they did not – have been used to explain the development of the basic *International Division of Labour* (IDL) that emerged from the nineteenth century and prevailed largely unaltered until the 1950s. In this system, shaped by colonial relations, the developing world was largely relegated to providing raw materials for the industrialized economies of the core (the United States, Western Europe). High-value manufactured goods were exchanged between the industrialized countries, and some were exported back to developing countries. This 'traditional' IDL was characterized chiefly by arms-length trade, and underpinned by a concentration of banking power and capital in the core countries.
- Second, from the 1950s on, a *New International Division of Labour* (NIDL) supposedly started to emerge. The NIDL thesis described the establishment by European, North American and Japanese TNCs of a global manufacturing system based on establishing labour-intensive export platforms in so-called 'newly industrializing economies' (NIE's). This process was driven by the need to maximize profits under conditions of heightened global competition. The system depended on new technologies that allowed production fragmentation, thereby creating tasks that could use semi-skilled or unskilled workers, often young and female, to provide cost-effective exports of goods back to the core. Coming from a neo-Marxist standpoint, the thesis was in part a critical examination of the consequences in these shifts for workers at both ends of this new spatial division of labour. Since its initial formulation by Fröbel, Heinrichs and Kreye (1980) the NIDL has been revealed to be a major simplification of recent global shifts in production. In effect, the NIDL describes just one specific form of IDL that has developed in a restricted range of labour-intensive manufacturing activities, rather than a new global mode of growth (see Fagan and Webber, 1999, for a full critique). Moreover, the NIDL thesis:

- ignores the continued importance of the 'traditional' IDL, for example in terms of continuing investment in raw materials ventures and agribusiness
- underplays continuing high levels of TNC investment between developed countries
- ignores the growing levels of outward investment from TNC's based in the NIEs
- overlooks investments aimed at serving growing domestic markets rather than for export (e.g. service sector FDI)
- overstates the significance of cheap labour and does not give enough emphasis to the importance of state policies (e.g. tax rates, labour market flexibility, market openness) and local capital (e.g. subcontractors) in NIEs for securing investment

- Third, the notion of the *Global Division of Labour and Power* (GDLP) (Mittelman, 2000) has been proposed to try and capture this sectoral and spatial complexity within the contemporary global economy. This approach emphasizes that there is no dominant mode of development:

> washing over or flattening diverse divisions of labour both in regions and industrial branches ... varied regional divisions of labour are emerging, tethered in different ways to global structures, each one engaged in unequal transactions with world centres of production and finance and presented with distinctive developmental possibilities. (p. 41)

Mittelman's formulation is important in three respects. First, it emphasizes the geographical variation within the world economy, and in particular, the supranational regional basis of many economic relationships. These geographies are far more intricate than the simple core-periphery model implicit in the IDL and NIDL approaches. Second, it incorporates the notion of variable power relations between different places in the global system that emerge as a result of these many and variegated divisions of labour. Third, it reflects the continued importance of the state in influencing the articulation of particular places with international divisions of labour.

- First, the *complexity* of international divisions of labour has been increasing. In some industries, such as clothing, a simple 'cloning' branch plant model of low cost overseas assembly persists. However, in other sectors, increasingly complex 'part-process' structures are developing, enabled by a range of developments in transport, communication and process technologies (see Box 4.4). This complexity is illustrated by Figure 4.3, which shows the production system for hard disk drives (HDD), a key component of the

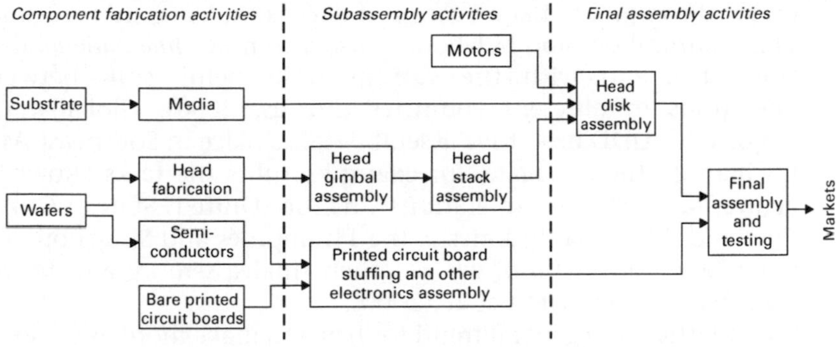

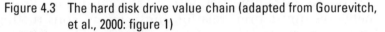

Figure 4.3 The hard disk drive value chain (adapted from Gourevitch, et al., 2000: figure 1)

Box 4.4 Technological change

The development of increasingly complex spatial divisions of labour has been enabled by a range of technological developments over the last several decades. Of particular importance are technologies that allow firms to overcome or limit the constraints of space and time. There are two key aspects to this. First, developments in *transport systems* have increased the potential to move materials, products, paperwork, people and other tangible entities between places. Of particular significance here have been the huge growth in commercial jet aircraft traffic since the 1960s, and the advent of containerization, which simplifies transhipment from one mode of transport to another. Second, developments in *communications systems* have facilitated the fast and secure transfer of a variety of forms of information. Here, the list of significant developments over the last 40 years is almost endless: satellite communication, fax, optic fibres, mobile telephones, etc. Of potentially revolutionary importance has been the recent emergence of a variety of networked computer systems that combine the power to both process and transmit information. Most obviously, the internet is fast becoming a global standard for instantaneous electronic communication and information sharing. By allowing the speedy circulation of goods, money, people and knowledge, developments in transport and communications technologies have played a critical role in facilitating the spatial separation of different parts of the production process. In addition, in some industries, developments in manufacturing *process technologies* have been significant in engendering new spatial divisions of labour in that through automation and standardization they have enabled the separation of the production process into elements with different skill requirements.

personal computers that we use most days as students and academics. This technical division of labour is also a complex *international* division of labour with the various tasks being split between operations in different countries (Borrus, 2000). Globally, the majority of HDD disk drive assembly takes place in Southeast Asia. US manufacturer Seagate, for example, splits the tasks shown in Figure 4.3 between operations in the United States, China, Thailand, Malaysia, Indonesia, the Philippines and Singapore, the latter being a key 'hub' where much final assembly and testing takes place (Gourevitch et al., 2000).

- Second, there is a general trend for firms to make more use of *externalized* networks (extra-firm relationships) as compared to *internalized* networks (intra-firm relationships) when constructing international spatial divisions of labour. This dynamic is reflected in the growing levels of international subcontracting and strategic alliances.
- Third, while spatial divisions of labour have always been fluid and shifting, they are arguably becoming *more dynamic* and liable to rapid change at the international level. This flexibility derives both from the use of certain technologies, and organizational forms that enable the fast spatial switching of productive capacity. In particular, this flexibility comes from the increased use of external networks described above. By using often spatially disparate networks of suppliers and subcontractors, firms can quickly switch contracts between different firms and places without incurring the sunk costs of moving production themselves.
- Fourth, international divisions of labour are becoming increasingly apparent in certain *service sectors*. Much analysis has focused on a small number of manufacturing sectors, namely clothing, footwear, electronics and cars, while overlooking the fact that service activities now account for over half of global FDI flows. Although much of this investment is of the market-serving 'cloned' variety, functional or 'part-process' divisions of labour are also now emerging, particularly for routine or back office functions. For example, some service firms now find it advantageous to conduct routine clerical functions in overseas sites – Ireland, India and the Caribbean are prime examples – with low cost land and labour. Jamaica, for example, exports services that revolve around the collection, transmission, storage, processing and display of information using information and communication technologies for largely US clients (see Figure 4.4).

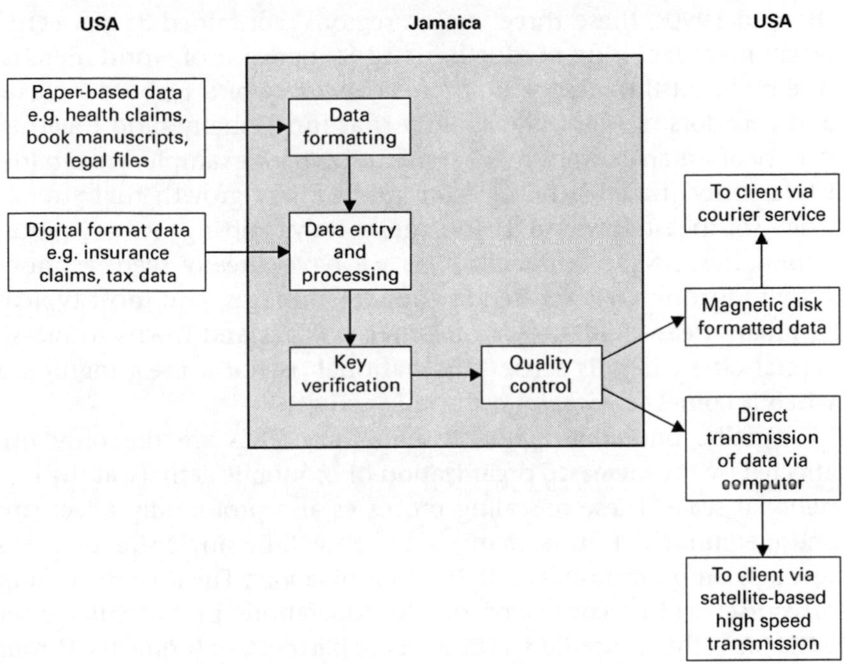

Figure 4.4 Stages in the international information processing industry (adapted from Mullings, 1999: figure 1)

There are three important caveats to accompany these general observations, however. First, these developments should not detract from what we have already said about the inherent 'localness' of capitalist production. Arguably, while the number of potential locations for economic activity to occur has – *theoretically* at least – burgeoned in the current era, the importance of place-specific attributes has arguably also grown as firms have become increasingly sensitized to inter-place differences. The specialized local agglomerations of economic activity described in Chapter 3 continue to be important characteristics of the economic landscape. What we are seeing, though, is the growing relative importance of various forms of *international* extra-local connections.

Second, and relatedly, the geography of the global economy that is produced by the aggregation of these myriad places and inter-places connections is extremely uneven. We can consider this unevenness at various scales (Dicken, 1998). At the macro scale, the world economic activity is dominated by the three triad zones of North America, Europe, and East and Southeast Asia. For example, in the mid-1990s, these three 'macro regions' contained 85 per cent of world manufacturing production and 80 per cent of world merchandise trade. At the meso scale there is a variety of transborder clusters and corridors of economic activity that functionally bind together a variety of urban centres. In Western Europe, for example, we can identify a concentrated band of economic activity growth that stretches from southeast England to northern Italy, cutting across national boundaries. At the micro scale, as we have already seen, economic activity is ultimately located in individual places, and most typically in urban areas of all sizes, from small villages and towns to massive capital cities. In this sense, the 'national' is not a meaningful scale when it comes to *mapping* economic activity.

Finally, one should not assume that TNCs are the only firms affected by the increased organization of economic activity at the international scale. These re-scaling processes also profoundly affect small and medium sized firms, many of which will be single site, or operate through simple intra-national divisions of labour. The fortunes of some, of course, will be connected to TNC operations in particular places, either directly, as suppliers, customers or partners, or indirectly, through consumption and reproduction practices. More generally, increasing numbers will have overseas suppliers, customers, and partners even if they do not have overseas operations themselves. Most importantly perhaps, firms of all sizes are open to the increased levels of foreign competition that accompany the upscaling of economic activity.

The upscaling of production can only occur where regulatory frameworks allow. Hence, large corporations in particular have been active in pushing states to create or participate in new supra-national regulatory institutions and agreements that facilitate their overseas expansion plans. This brings us to the next section.

Re/scaling regulation and the state

We now turn to the scaling of regulatory processes, and in particular to those changes associated with what Swyngedouw (1997) calls the 're-scaling of the state'. In the three decades after World War Two, the national state in advanced capitalist societies grew to occupy a central role in many areas of economic and social life. Elsewhere, many other countries were led by strong national states (as in the former communist bloc or the so-called Newly Industrializing Countries of South America and Southeast Asia), while others languished under a weak or divided state apparatus (as in postwar Ethiopia and Mali). The last 20 years, however, have seen a profound rolling back of state intervention in many 'first' and 'second' world capitalist societies and the emergence of a range of new state and quasi-state institutional forms and regulatory mechanisms. These changes have been part of the emergence of what's called 'neoliberalism' – led by the United States as the world's economic superpower – which promulgates the value of the free market as the disciplining mechanism for many fields of socio-economic activity (Box 4.5).

Box 4.5 The 'roll-out' of neoliberalism

It is possible to think of neoliberalism as 'the extension of market rule and disciplines, principally by means of state power' (Tickell and Peck 2003). So we can think of forms of deregulation and liberalization in areas such as trade, finance and labour markets as examples of neoliberal policies. However, the emergence of neoliberalism does not mean less activity for nation states: rather it means changes in the way they regulate the economy and society, implying the blurring of the boundaries between the state and the market. Since its emergence in the early 1980s neoliberalism has become the dominant state form in Western Europe and North America. Below a distinction is made between 'roll back' neoliberalism, where nation states dismantled existing ways of doing things, and 'roll out' neoliberalism where new systems were put in place.

	Roll-back neoliberalism	Roll-out neoliberalism
Periodization	1980s/early 1990s	Late 1980s–
Dominant discourses	Small government/ deregulation	Paternal state/free economy
Key institutions	Governing parties	State elites
Mode of political rationality	Ideological project	Technocratic management
Sources of resistance	Organized labour	Cyberactivists
Spaces of resistance	British cities and coalfields, North American rustbelt	Anti-globalization confrontations, France, Malaysia
Service delivery	Privatization	Marketization
Labour regulation	Mass unemployment	Full employability
Geographic heartlands	London and Washington, DC	Brussels, London and Washington, DC
Geographical frontiers	Brussels	Paris, Berlin, Hong Kong, Singapore, Johannesburg
Scalar constitution	National	Glocal

Source: adapted from Tickell and Peck (2003)

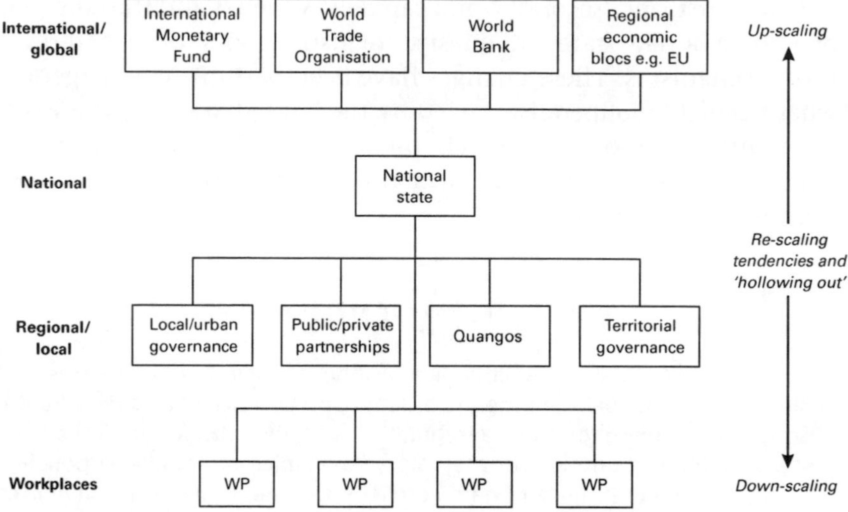

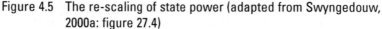

Figure 4.5 The re-scaling of state power (adapted from Swyngedouw, 2000a: figure 27.4)
From Clark G. et al., 2000, *The Handbook of Economic Geography*. Reprinted by permission of Oxford University Press.

More specifically, these changes in the way the states intervene can be read as a response to pressures 'from above' and 'from below'. Pressures 'from above' have come from the transnationalization of the economy described in the previous sub-section and the growing significance of supranational regulatory bodies such as the International Monetary Fund (IMF). Pressures 'from below' have been generated by a range of regionalist and localist movements within countries, driven by a range of cultural, economic and political motives. For example, successful regions may seek to limit fiscal transfers to less successful areas (in effect, a form of intranational place competition). These twin pressures have engendered tendencies to shift regulatory practices 'upward' to the supranational level, and 'downward' to the regional level (see Figure 4.5). While the presence of supranational and subnational power structures is far from new, the range of such organizations and the power that they hold is now unprecedented. At the international scale we can recognize the considerable influence of the IMF, World Bank and World Trade Organization (WTO) (see Table 4.1) along with a wide range of macro-regional regulatory forms of which the European Union is the most advanced case (see Box 4.6). Importantly, in recent times these supranational bodies have generally pursued a deregulatory agenda that has fuelled the growing levels

TABLE 4.1 Key supranational regulatory institutions

Institution	Vital statistics	Key roles
World Trade Organization (WTO)	• Founded: 1995, replacing GATT[1] (1947–1994) • HQ: Geneva • Staff: 550 • Member countries: 144 (as of 1/1/02) • More than 90% of world trade is covered by the WTO framework	International trade organization responsible for the regulation of global trade – rules derived by negotiation among member states *Main functions:* • administers WTO agreements • forum for trade negotiations • handles trade disputes • monitors national trade policies • technical assistance and training for developing countries • cooperation with other international organizations
The World Bank	• Founded: 1944 • HQ: Washington DC • Staff: 10,000 + • Member countries: 184 • Provided US$17.3 bn of loans in 2001	Provider of development assistance – in terms of both knowledge and finance – to over 100 developing countries *Has activities in the following areas:* • basic health and education provision • social development and poverty reduction • public service provision • environmental protection • private business development • macroeconomic reform
International Monetary Fund (IMF)	• Founded: 1944 • HQ: Washington DC • Staff: 2,650 • Member countries: 184 • Has outstanding credits and loans to 88 countries of US$88bn (as of 30/6/02)	Central institution of the international monetary system, charged with promoting stability and efficiency *Main areas of activity:* • monitors national economic policies and performance • gives policy advice to members • lends to members with balance of payments problems • provides governments with technical assistance and training

[1]General Agreement on Tariffs and Trade
Sources: www.wto.org, www.worldbank.org, www.imf.org

of international production, investment and trade described in the previous section. Equally, in many countries we see (subnational) regional forms of governance gaining more powers, particularly in areas such as economic development and training. Countries vary in the degree to which this trend has occurred, but even a traditionally highly centralized state such as the UK has recently instigated a Scottish Parliament, elected assemblies in Wales and Northern Ireland, and Regional Development Agencies across England.

Box 4.6 Types of regional economic integration

Regional economic blocs are a significant addition to the regulatory architecture of the global economic system. While the initial stimulus for such formations comes from a desire to reduce barriers to trade, more integrated forms can also emerge. Following Dicken (1998), we can identify four key types of bloc:

- **The free-trade area:** trade restrictions between member states are removed, but states retain their individual trading arrangements with non-members e.g. the North American Free Trade Agreement (NAFTA) between Canada, the USA and Mexico since 1994.
- **The customs union:** members operate a free trade agreement between themselves and have a common trade policy for non-members e.g. the MERCOSUR customs union between Argentina, Brazil, Paraguay and Uruguay since 1995.
- **The common market:** has the characteristics of a customs union, but in addition, allows the free movement of factors of production (e.g. capital and labour) between members e.g. the Single European Community Market since 1993.
- **The economic union:** harmonization and supranational control of economic policies e.g. only the European Union comes anywhere close to this form as evidenced by the adoption of the euro as the sole currency of several member states in 2002.

Three further points should be made. First, numerically, the vast majority of agreements fall under the first two headings (free-trade areas and customs unions). Second, economic blocs may develop over time and 'move down' this list, as has been the case with what is now the European Union. Third, it is important to recognize that all these regional economic forms are initiated by, and derive legitimation from, their member states.

In addition to these re-scaling dynamics, there has been a simultaneous shift in emphasis from processes of *government* to *governance*, with regulatory capacity being 'hollowed-out' to quasi-state institutions at the regional, national and even international scales (Jessop, 1993). For critical commentators this shift has seen democratic, elected forms of government be replaced by less accountable, opaque institutions dominated, in many cases, by business elites. A UK example would be the Urban Development Corporations (UDCs), limited-life quasi-state institutions established in the 1980s (and wound up in the late 1990s) in order to transform selected old industrial areas in the major cities into environments in which private capital would invest. UDCs were non-elected bodies appointed and generously funded by central government. For proponents of UDCs, they were successful institutions that transformed the physical environment and created thousands of jobs in areas blighted by high unemployment and poverty. For critics, they were authoritarian bodies driven by private sector interests – at the expense of local community representation – that created relatively few high quality jobs in return for the level of investment.

Three further points can be made about this regulatory re-scaling. First, what we are seeing here is the changing character of state involvement in political-economic regulation, not its withdrawal. Indeed, the processes of re-scaling are themselves the result of conscious policy decisions taken by states. Second, the process of 'hollowing-out' is arguably a Western European phenomenon. Many non-Western governments, for example, never had anywhere like the range of powers that several Western governments are now re-scaling. Moreover, 'hollowing out' does not mean less of a role for nation-states, but marks a qualitative transformation in it regulatory capacity. Third, there is enormous variation within this multi-scalar governance system as to the power that different states retain. At one end of the spectrum, the United States is extremely powerful, and is a dominant force within institutions such as the WTO and IMF. At the other end, many developing countries are so dependent on outside assistance that they are forced to adopt regulatory regimes imposed 'from above' by institutions such as the IMF.

Re-scaling workers and unions

Here we will focus largely on the scaling of trade unions as the pre-eminent independent form of worker organization. Trade unions have traditionally been – and still are – organized at three key scales, namely the workplace, the 'local' and the national (see Figure 4.6). Most workers' union involvement is enacted through the workplace, many of which are then aggregated together into some form of

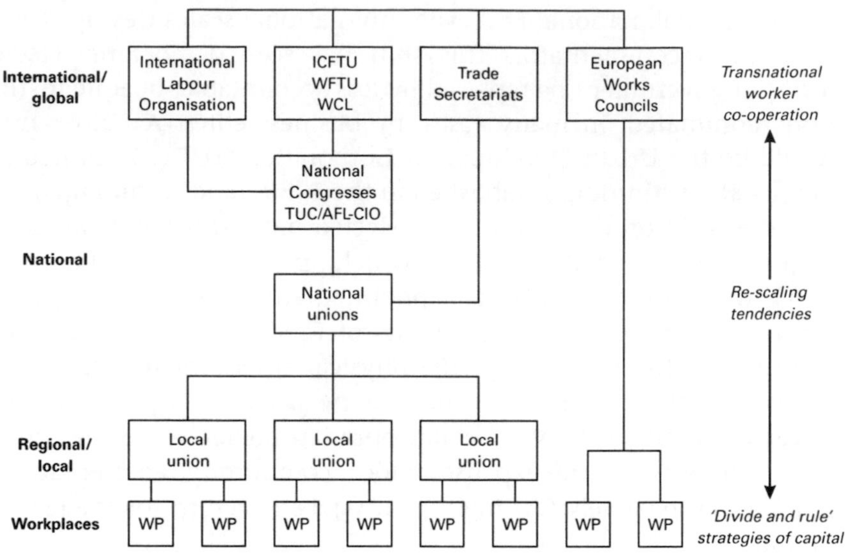

Figure 4.6 The re-scaling of trades unions

sub-national union 'local' (for example, within a city, county or region). In turn, in most cases, these locals form part of a national organization. Even this national-local-workplace hierarchy is a simplification of a complex reality, however (see Figure 4.7), for as Martin et al. (1996: 15) describe,

> The institutions and processes of industrial relations operate at a variety of scales ... most unions are organized into different geographical levels, for example into regional, district and local branch divisions, so that this itself imparts a material spatial structure into the industrial relations system. Furthermore, varying degrees of autonomy of orientation and operation, of custom and convention, may exist at these different levels.

As the quote suggests, it is very difficult to generalize about at which level in this hierarchy different powers reside, as autonomy will vary along a number of dimensions. Most importantly, the relative importance of these different scales of union activity will vary between different national economic contexts. During the postwar boom in the UK, for example, equally high levels of national collective bargaining mirrored high levels of union membership. Collective bargaining is a process whereby different workers in different places are represented by one relevant union body that negotiates at the national level on behalf of all its members with the relevant employers. During the

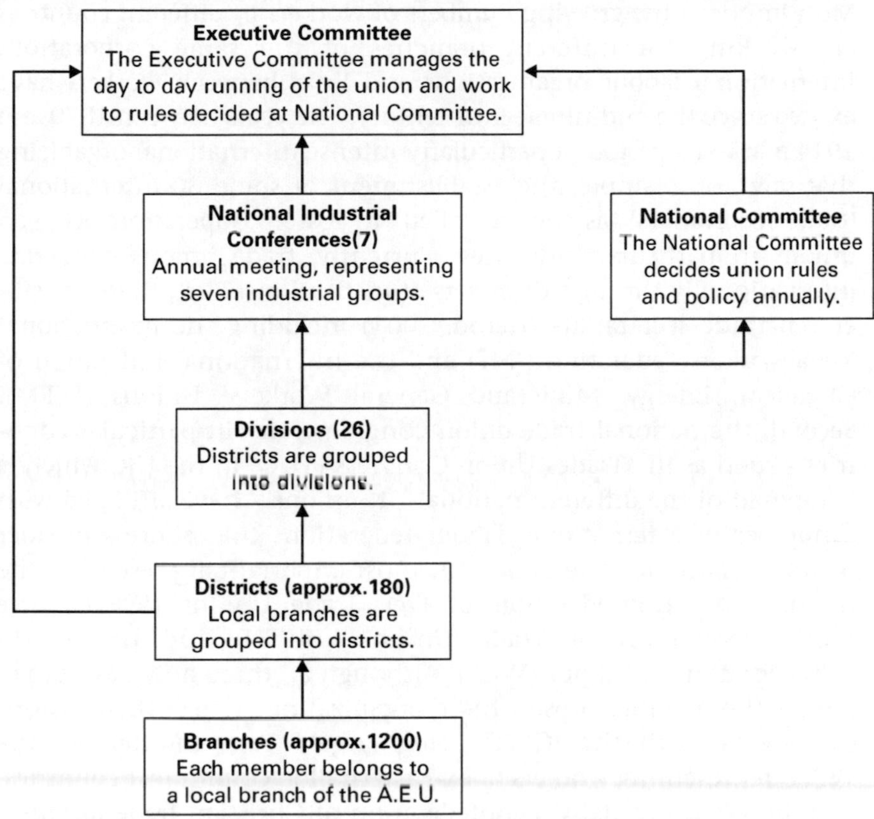

Figure 4.7 The geographical and organizational structure of the AEU, C. 1990 (adapted from Martin et al., 1996: figure 6.3)

1980s, however, strict new legislation was enacted by Margaret Thatcher's Conservative Party administration, with the result that the locus of collective bargaining swung back towards the local and workplace scales. Equally, there will be important sectoral and union-level differences in the 'scaling' of worker–employer bargaining. In many contexts, however, the trend is towards a 'localizing' of bargaining activity as employers and/or governments increasingly put pressure on unions to decentralize the geographical scale at which bargaining takes place. Many national agreements on pay and conditions were dismantled in the United States in the 1980s, for example, giving firms greater opportunity to play plants and locations off against one another.

'Above' the national scale, the profound scalar restructuring of the global economy over the last few decades has created both challenges and new opportunities for transnational trade union organization.

Most importantly, growing numbers of workers in different countries are working for different branches of the same corporation. International labour organizations and links between workers have existed since the mid-nineteenth century. The years between 1870 and 1914 marked a period of particularly intense international organizing that saw, for example, the establishment of some 30 international 'trade secretariats' (as they're called) to foster cooperation between unions in particular industries. Currently, trade unions cooperate internationally through three sets of institutions. First, there are still eleven trade secretariats (Herod, 2001) including the International Metalworkers Federation (IMF) and the International Federation of Chemical, Energy, Mine and General Workers' Unions (ICEM). Second, the national trade union conglomerates in particular countries – such as the Trades Union Congress (TUC) in the UK, which is composed of the different national UK unions – have affiliated with a number of international labour federations that address broader issues than do the secretariats. The most important at present are the International Confederation of Free Trade Unions (ICFTU), the World Federation of Trade Unions (WFTU) and the World Confederation of Labour (WCL). Although all three now have broad remits, the memberships of these organizations reflect their historical origins, with the ICFTU being largely Western-oriented, the WFTU representing workers in ex-Eastern Bloc/communist countries, and the WCL – initially a confederation of Christian trade unions – having a strong presence in the developing world. Third, trade unions have pursued labour and human rights issues through a range of inter-governmental labour institutions, most notably the International Labour Organization (ILO), an agency of the United Nations (UN).

Comparing Figures 4.2, 4.5 and 4.6 may seem to suggest a neat symmetry in which workers have a range of organizations to counter the parallel re-scaling of economic forces and regulatory dynamics. In reality, however, the actual *effectiveness* of international labour organizations has been extremely uneven. Many lack both material resources and institutional influence. Workers in different countries remain divided by a broad range of capitalist, non-capitalist and regulatory differences that pose real barriers to achieving international solidarity. Trade unions for the most part remain profoundly *national* organizations, which as Jane Wills (quoted in Hudson, 2001: 106) observes 'think and act nationally, despite acknowledging the importance of the international scale'. However, new forms of less-formalized international connection between workplaces and unions are emerging – such as so-called 'social movement unionism' or European Works Councils – often stimulated by the restrictively hierarchical nature of many of the international labour organizations described above (Herod, 2000a). We will return to these initiatives in Chapter 8.

BRINGING IT ALL TOGETHER? LOCAL LABOUR CONTROL REGIMES

We now have an appreciation of how what happens within particular places may be profoundly influenced by a variety of extra-local connections. Equally, we have seen the importance of the scaling – and indeed re-scaling – of these connections at multiple levels. But what do these insights mean for our analysis in Chapter 3, which clearly showed how workers, firms, and regulatory mechanisms are all very much 'placed', in the sense that they ultimately interact locally and in the workplace? The *local labour control regime* (LLCR) is an integrative concept that can help us to bring together these different aspects. In short, it brings us back to the notion of places as unique, yet simultaneously open, porous and interdependent. It explicitly recognizes that the nature of worker–employer relations in a particular locality is simultaneously determined by local *and* extra-local dynamics at a variety of scales up to, and including, the global. The development of these regimes is driven by one of the basic contradictions of capitalism, namely that which exists between the *potential* spatial mobility of many firms, and their associated bargaining power, and the need for firms to extract profits from concrete investments in particular localities, which requires a certain measure of regime stability. The term 'regime' is used here to denote a place-specific system of social relations, norms, rules and habits inside and outside the workplace through which employers make their profits.

The clearest formulation of this concept is offered by the economic geographer Andrew Jonas (1996: 325), who describes an LLCR as a 'historically contingent and territorially embedded set of mechanisms which coordinate the reciprocities between production, work, consumption and labour reproduction within a local labour market'. Importantly, this formulation extends well beyond the workplace to incorporate the intimately connected domains of consumption (housing, recreation, household consumption, etc.) and reproduction (education, training, health care, welfare, etc.) that we identified in Chapter 2. The notion of 'control', then, is construed here in a broad sense, rather than just in terms of workplace measures to improve efficiency or productivity. It naturally follows that the full range of worker, household, firm, civil society, state and quasi-state institutions are – or at least may be – involved in shaping the LLCR.

Thus, on the one hand, a LLCR constitutes the unique, place-specific relations between firms, workers, unions and regulatory institutions that enable labour to be integrated into production. At the same time, it is defined and shaped by multi-scalar extra-local processes, and the broad range of competitive and cooperative extra-local worker, employer and regulatory connections identified in this chapter will influence its character. Put another way, every local regime is 'nested' within labour control regimes operating at larger scales that will influence, *but not determine*, the nature of employer–worker relations in that particular locality. As we have seen, of particular importance is the way in which the regime is integrated into production systems and corporate spatial divisions of labour, organized at the wider national and international scales that *actively seek* to take advantage of the differences between various regimes. A point that follows on from this 'scaling' of the local control regime is that actors and institutions organized at other spatial scales will have an influence on processes within it. Most obviously, multi-plant firms will take decisions that reflect operations in, and across, a variety of regimes. Equally, national state and other supra-local regulatory institutions will institute a range of policies that impinge upon and shape the LLCR.

The exact nature and operation of a regime will largely depend on the differential power relations between the constituent groups, as they play out *in situ*. As we saw in Chapter 2, we can think of power relations as operating on two different, though interrelated, levels. First, there are a variety of 'structural constraints' that may impinge upon an LLCR in a number of ways. For instance, the realities of global competition in certain sectors may make the imposition of certain terms and conditions for workers inevitable. Equally, states may impose regulatory constraints that circumscribe the potential strategies of workers and firms in a number of ways. Second, there is a wide range of 'power repertoires' through which workers, firms and regulatory bodies can exercise their agency within the local regime (refer back to Box 2.2). This in turn may reflect the form and success of alliances that develop between actors in the LLCR in varying combinations. In some cases, as we shall see in Chapter 6, formal and informal alliances between businesses and governments – both local and/or national – desperate to attract or retain investment may serve to dis-empower labour in certain localities. However this is not always so: in other situations, more progressive local coalitions may develop to maintain or improve worker conditions (the argument of Chapter 6).

It is also important to consider hegemonic discourses circulating within the LLCR that may augment the power structures and repertoires outlined above (Coe and Kelly, 2000). There are three aspects to this. First, influential discourses are mobilized – such as those concerning globalization – that shape a particular understanding of the 'external' economy at the national and global scales. These 'scalar discourses' form an important part of power repertoires as groups displace responsibility for particular processes to the geographic scale – 'upwards' or 'downwards' – that best suits them. Second, these representations of the external relations of the LLCR may be used to justify and operationalize particular strategies *within* the locality, for example, reducing labour costs to remain 'competitive' in international terms. Third, representations circulating within the local regime can serve to validate and prioritize certain segments of the labour market – such as 'knowledge' workers – over others, thereby potentially creating problems for maintaining its stability.

As a result of the complex intermingling of local and extra-local influences, a regime, although characterized by an element of stability 'is not a static and fixed object but rather a fluid and dynamic set of social relations and power structures which are continuously reproduced and/or transformed by the forces of domination, control, repression and resistance operating at a variety of scales' (Jonas, 1996: 329). Although many analyses place a heavy emphasis on firms as agents of change, in reality developments may be instigated locally or extra-locally, and by worker, firm or regulatory institutions, depending on the context.

The global capitalist economy is characterized by a bewildering variety of different LLCRs. Even within the same country and the same industry, the differences may be profound. In the computer industry, the highly flexible innovative labour regime of Silicon Valley is supposedly very different from the large-firm dominated regime found in the Route 128 area near Boston (Saxenian, 1994). A corollary of this great variability is that the problems and possibilities facing workers will differ dramatically between different places. After all, it is only certain kinds of businesses that benefit from inter-locality flows of investment and aggressive spatial restructuring. Other businesses serve to benefit far more from becoming progressively involved in the LLCR and community development (Jonas, 1996).

WORKERS, GEOGRAPHICAL POLITICS AND GEOGRAPHICAL DILEMMAS

The geography of worker politics

By now it should be clear that geography is political. It is both a resource for workers and a threat, just as it can be for the firms that employ them and the institutions that regulate them. To comprehend the complex ties between place and space, which is to say the connections among scales, is to be equipped with a geographical imagination. It's an imagination, as we will see in the rest of the book, that many wage-workers possess in abundance, even while others remain locked into thinking and acting at one scale only. In this penultimate section of the chapter we want to offer some systematic reflections on how geography both structures and complicates the options available to wage-workers in any given situation.

In this and the previous two chapters we've depicted a world in which wage-workers are involved in social relations with employers and regulatory institutions. These are relations of conflict and co-operation, of control and interdependency, of fixity and motion. They are profoundly geographical relations that are at once expressed in, and shaped by, place and space: they are, in short, scaled in complex ways. In light of all this, any given group of workers in any given place can pursue one or all of a combination of the following inter-linked types of *geographical strategies* in order to enhance pay, conditions or rights of themselves, fellow workers or related non-working groups elsewhere:

1 **Act locally for local needs and wants** – here workers in a given place undertake resolutely local scale action to serve locally specific worker, employer, dependent, retiree or unemployed needs (for example, by 'keeping out' foreign migrants, competing for local jobs, or accepting pay cuts to stop a company relocating elsewhere).
2 **Act locally for the non-local needs and wants of others** – here workers, in a given place undertake resolutely local action to serve the needs of workers, etc. in other places (for example, by refusing to buy commodities made in distant places where oppressive labour conditions prevail).
3 **Act translocally for local needs and wants** – here workers in a given place undertake translocal actions in order to serve locally specific worker, employer, dependent, retiree or unemployed needs (for example, as in the Liverpool dockers' example in Chapter 1,

where Terry Teague helped lead an international docker campaign against the Merseyside Docks and Harbour Company; as when labour migrants move elsewhere in order to send remittances back to their families).

4 **Act translocally for the non-local needs and wants of others** – here different workers in myriad places undertake coordinated inter-place action in order to improve the pay and conditions of many or all of the workers involved (for example, as in campaigns to institute a global minimum wage or to institute a universal charter of worker rights).

Note that it is easier to distinguish these geographies of worker action analytically than it is practically. We will be exploring examples of each in the chapters to follow. For now we simply want to emphasize three things. The first is that each of these strategies begins and ends in place. Even translocal actions – like those undertaken through international trade unionbodies – have a local dimension, since they are ultimately based on local level practices. As Harvey (1995) argues, all labour struggles start and finish with 'militant particularisms'. Second, for wage-workers, even though place might not be the only scale of actual or potential action, there can be real tensions between workers' needs in one place and those in another. These tensions can be serious barriers to achieving inter-place worker cooperation over wages, rights and working conditions (see Chapter 9). Finally, wage-workers can pursue one or more of the four geographical strategies in relation to the five principal types of politics identified at the end of Chapter 2. This inserts tremendous complexity into the politics of worker action.

Geographical dilemmas

This brings us to a key problem for contemporary workers that will be made much of in Chapter 9 but which runs through all the chapters to come: the problem of *geographical dilemmas*. A dilemma is a situation where pursuing one seemingly logical or sensible course of action can, in fact, have negative consequences for some of those involved. In a capitalist world, workers face *geographical* dilemmas because what might make sense for them at one geographical scale may have unfortunate consequences at other scales. For example, where workers in one place compete with those elsewhere for jobs and investment they are serving local interests but at the expense of the livelihoods of people elsewhere with similar needs and wants. Alternatively, where workers in one place join a translocal campaign to help workers elsewhere – as dockers Cees Cortie and Bruce Robinson did on behalf of Terry Teague and colleagues – they may have to sacrifice their personal and local interests in the process. The crux of these and other paradoxes is the geographical scale at which workers define their interests and their loyalties (Harvey, 1995). It's easier for workers to be loyal to those who they see and work with everyday, but much harder to feel ties of solidarity with wage labourers elsewhere. As Harvey (2000: 79) notes, 'we all too often lock ourselves into one ... geographical scale of thinking, treating the difference at that scale as *the* fundamental line of political cleavage'. This tendency of workers, in the first instance, to define their interests at the local scale is exacerbated by the fact of uneven geographical development highlighted earlier in this chapter. For example, workers in poor parts of the world are hardly likely to support the efforts of those in the developed world to improve already relatively good pay and conditions. So it's not just that workers may be tempted to put local interests first, it's also that the very *nature* of local interests varies depending on the specifics of local industry, local standards of living, local living wages and so on.

This raises the thorny issue of justice for wage-workers. We mentioned in the Preface that we've written this book as authors seeking to be sensitive to injustices done to workers in different places. In the most general sense justice concerns the fairness or rightness of how a person is treated by others (we do not consider environmental in/justice in this book, that is, the propriety of how people treat the non-human world). Justice is therefore a social phenomenon because it involves inter-personal and inter-group relations. Injustice is frequently the result of unequal power relations between individuals and groups. In the formal sense, standards of justice are enshrined in labour and civic law and are usually upheld by the judicial arm of states. In a less formal sense, tacit understandings of what's 'fair' and 'unfair' treatment of workers are thrashed out between specific workers, employers and regulatory institutions in different geographical contexts. The struggle against injustice to workers is thus a daily one, with tradeunions typically the most forceful advocates of worker rights at the workplace, firm and sectoral levels. Implicit in all concepts of in/justice is a standard to define what counts as 'good'/'bad' or 'acceptable'/'unacceptable' practice. This standard has a critical function (to expose injustice) and a constructive function (to establish a clear marker for justice).

It's not hard to see that in/justice is not just social; it's also geographical. Thus, we might propose a common standard of justice for workers in different places and/or at the very least be vigilant about yawning geographical variations in how workers are treated. Yet this begs the questions: what constitutes 'injustice'?; and by what criteria do we define a 'just' situation for workers? These are difficult questions to answer, for two reasons. First, they cannot be responded to in absolute terms. Though it might be tempting to offer fixed, universal definitions of in/justice because they give us a clear and widely applicable socio-geographic benchmark, the fact remains that in/justice is and must always be culturally defined. Thus what might be considered unjust treatment of workers in Sweden might scarcely be deemed important in Indonesia. Second, this means that we have to reckon with diverse conceptions of what's un/just existing worldwide. This diversity cannot be wished away. The challenge is to acknowledge and respect it, while arguing for more socially and geographically universal conceptions of justice that can expose the worst excesses of worker exploitation. In the words of geographer Don Mitchell (2000: 293), any robust notion of worker in/justice must be 'fully situated within a politics of scale'. Thus, although the idea that all workers should not be made to work in conditions that actively damage their health is culturally defined – it's what geographer Wolfgang Natter (1995) calls a 'contingent universal' – we would argue that it deserves worldwide applicability as a standard for in/justice. Global standards of this kind, designed to uphold certain basic rights for workers, do not erase other, culturally specific criteria of justice at the sub-global scale. Rather, they place a 'floor' beneath them and can provide different workers in different places with some common goals and aspirations. To revisit the questions posed above, then, the appropriate thing for labour analysts to ask is not 'What is just/unjust for workers?' but rather 'What *kinds* of justice is appropriate for *which* workers at *what* geographical scales?' This latter question opens up a rich and complex terrain that is attuned to the very real differences – notwithstanding the similarities – among workers worldwide (a terrain we explore in Chapter 9).

Finally, this brief discussion of in/justice brings us to our normative standpoint as authors of this book. In the Preface we announced our Marxian and institutionalist predilections, ones that place us on the left of the political spectrum. But we wish to depart in one significant sense from the tenor of Left discussions of wage-workers typical in human geography, sociology and other disciplines. These discussions frequently use terms like 'progressive' and 'regressive' in a rather inflexible, absolutist way when passing judgement on worker struggles. For example, local efforts by workers to compete with other places for investment are often dubbed 'regressive' because they undermine wider, inter-place working-class solidarities. However, it seems to us wrong to pass judgement in this way. Terms like 'progressive' and 'regressive' are relative. That is, they can only be applied *contextually* and with reference to the workers involved. As we will see in the rest of the book, a worker action that is 'progressive' for some is simultaneously 'regressive' for others. The paradoxes referred to above cannot be sidestepped and this is why it is difficult for labour analysts to make grandiose claims about the rightness or wrongness of any given worker action at any given scale/s.

SUMMARY AND PROSPECT

Readers should now have the theoretical equipment to make sophisticated sense of the opportunities and constraints that condition what happens to wage-workers in the modern world and how workers respond. To summarize: if we combine the insights of this and the previous chapters it's clear that wage workers are enmeshed in what Piven and Cloward (2000: 414) describe as 'a tangle of complex and concrete relations'. This tangle is simultaneously social and geographical. It means that workers in our capitalist world are, at one and the same time, intensely differentiated in a socio-geographic sense *and* intensely connected. To be sure, the nature and intensity of the connections are by no means even. But they are quite real. So what implications does this combination of difference and interdependence have for different groups of workers? How can the theoretical arguments presented in this Part of the book help us understand the condition of wage-labour in the contemporary world?

The remaining chapters will answer these questions, using case studies drawn from around the globe. They are structured as follows. In Part Two we consider the struggles of wage-workers at the local level. Chapter 5 starts by showing the various ways in which businesses and certain regulatory institutions (especially those of the state) conspire to put workers on the back-foot. This chapter, if you like, shows the extent to which the 'globalization script' criticized in Chapter 1 is true vis-à-vis workers. That is, it explores those cases where locally-based workers are seemingly powerless in the face of inter-place competition and firm mobility. By contrast, Chapter 6 explores the ways that proactive actions by workers, employers and local institutions can retain or attract jobs and investment to different places. After these two chapters on the local and sub-local scales, Part

Three switches attention to translocal worker action. Chapter 7 examines workers on the move, that is, labour migrants. Chapter 8 then looks at contemporary efforts to construct translocal webs of worker co-operation through international unions and other bodies. Part Four then draws things together by exploring the quest for worker justice and the geographical dilemmas involved.

Part Two

LABOUR IN PLACE

five

Re/placing Labour

A report published by the Oxfam-led Clean Clothes Campaign (Connor, 2002) drew the following conclusions from an investigation into the living and working conditions of Nike and Adidas workers in West Java, Indonesia. First, with wages as low as US$2 per day (including working substantial overtime), many workers were living in extreme poverty. Those with children had either to send them to distant villages to be raised by relatives or enter into debt in order to meet their basic needs. Second, workers' freedom of association was strictly curtailed. Many clearly feared that active union involvement would lead to them being dismissed, imprisoned or physically assaulted. Third, workers were subjected to harsh physical and psychological pressures in the workplace, being routinely shouted at and humiliated, and made to work in dangerous conditions. One woman interviewed in the report was earning US$45 per month making foot ball boots for Nike, but only after being forced to work the maximum amount of overtime each week. She lived alone in a tiny room and sent much of her wage – after rent – back to her family. When asked whether she had thought about doing other jobs, her answer was simple: 'Tell me. What else can I do?' (Connor, 2002: 30).

This vignette reveals one disturbing and demeaning local 'world of work' under global capitalism. Its a world where developed nation TNC's such as Nike and Adidas constantly switch, or at least threaten to switch, production between low-cost subcontractors in a variety of developing world locations. People living in already deleterious

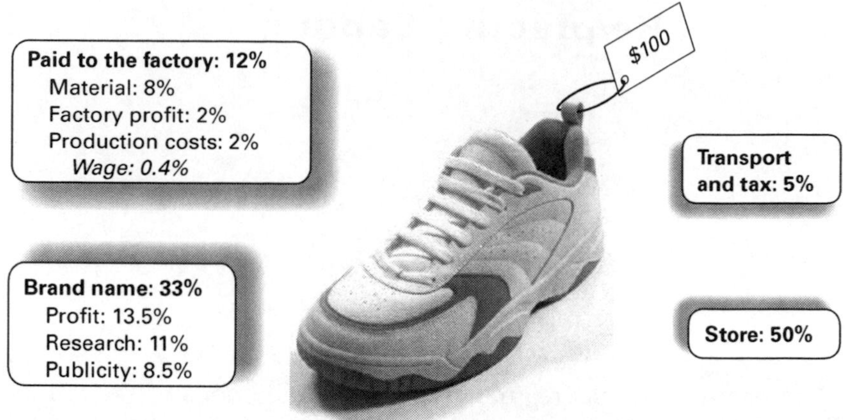

Figure 5.1 The price make up of a $100 sports shoe made in Indonesia (adapted from www.cleanclothes.org/campaign/shoe.html)

conditions seemingly have little choice but to enter waged work in order to try and enhance their position. In turn, by exercising a variety of often brutal power repertoires, and in particular by bringing workers in different places into direct competition with one another (a process commonly know as 'whipsawing'), employers are able to keep labour costs to the bare minimum. In the case of the sports shoe industry, for example, this has been done so effectively that manufacturing labour costs may account for just 0.4 per cent of the total retail price of a US$100 shoe (see Figure 5.1), while at the same time firms claim that securing low wage costs are critical to their overall competitiveness (Schoenberger, 1998).

This is just one of many local worlds of work that coexist – and, of course, cooperate and compete – within the structures of global capitalism. The sports shoe labourers in Indonesia themselves are intimately linked to the worlds of work of highly-paid purchasing, publicity, marketing and brand specialists based in headquarters in the United States and Germany, and of armies of relatively lowly-paid part-time and temporary retail staff in sports shops around the world. While the globalization myths mentioned earlier in the book purport that global interdependence is having a range of regressive impacts on workers around the world, in reality the consequences for labour are many, complex and often contradictory. The aim of this chapter is to explore the various ways in which businesses and regulatory institutions are increasingly 'squeezing' concessions from labour in an era of the increased geographic mobility of capital. The central argument, however, will be that both these strategies and their outcomes are many and geographically varied. The chapter proceeds in three stages. First, we recount, and then critically evaluate, the understanding of global capitalism (or what some call globalization) that emphasizes the deleterious impacts of capital mobility on place-based workers. Second, recognizing that there are elements of truth to this argument, we delimit the range of labour control mechanisms through which firms and states are eroding the pay and conditions of workers in the contemporary context. Third, we use three case studies to illustrate the tremendous intra- and inter-place variation in how these various mechanisms are combined and implemented.

AN EMERGING GLOBAL REGIME OF 'HEGEMONIC DESPOTISM'?

Let us start by reviewing a general account of global capitalism that is currently put forward by analysts of various political leanings, left and right. This account corresponds to the labour-related myths of globalization discussed in Chapter 1. The basic argument goes as follows. A defining characteristic of contemporary capitalism is the *relative* geographic mobility of capital compared to that of labour, which becomes the source of the 'asymmetrical' power of the former over the latter. This mobility can be seen to form the basis of a new 'global politics of production' (Jonas, 1997) in which the persistent threat of relocation becomes a powerful tool for employers when determining wages and benefits, contracts, investment strategies and the like with employees. The *potential* mobility of businesses on a global scale thus becomes a critical weapon for corporations negotiating with workers and local interest groups. Workers are increasingly pressured into having to defend their interests in particular places, and in the process may have to compete directly with workers in other places to keep workplaces operating and secure jobs. This competition leads to a progressive 'ratcheting-down' of labour terms and conditions as mobile capital seeks out the best rates of return. These conditions have led some to suggest that we are now seeing a neoliberal regime of 'hegemonic despotism' in which labour is subordinated to business in a process of competitive undermining of labour standards and salaries (Burawoy, 1985). This competition can occur at the international and/or subnational scales. Britain and France may compete, for example, to host a new Japanese car manufacturing plant. Equally, locations within those countries may also be drawn into competition with one another in order to attract much needed investment. Large quantities of public money may be invested into specific bids to secure investment, or place marketing more generally. For example, in the early 1990s, the state of South Carolina gave BMW US$300 million worth of subsidies to secure a new car plant (Brecher and Costello, 1998). The hegemonic regime as a whole is knowingly shaped and sustained at the global scale by a powerful combination of political and corporate elites who together constitute what might be termed a 'transnational capitalist class'.

There is more than a grain of truth to this account. The geographic mobility of productive capital *has* certainly increased over the last few decades. Synthesizing our arguments from Chapter 4, we can identify five inter-linked contributory factors here:

- **Technological developments:** dramatic improvements in productive, transport and communications technologies have facilitated increasingly complex geographic arrangements of production.
- **Deregulation:** the lowering of regulatory barriers to international trade and foreign direct investment increasingly enables the transfer of both finished goods and factors of production (materials, technologies, people, and capital).
- **Increased competition:** the intensity of inter-place competition has increased, fuelled by the above two factors. Technologies, for example, have contributed to productivity growth that has often outstripped the expansion of markets, and thereby enhanced cost-competition. Deregulation may both increase competition in a firm's home market, and increase opportunities abroad.
- **Increased mobility of financial capital:** again aided by deregulation and technological developments, an international financial system has developed that can rapidly circulate huge quantities of capital in a search for profitable investments.
- **Pre-existing patterns of geographic uneven development:** a final stimulus to the mobility of capital is the profound and sustained spatial variation of the global economy, at all spatial scales. Of particular importance in the contemporary era are inter-place variations in labour costs, productivity, skills and levels of union membership and militancy.

However, the argument we wish to pursue in the remainder of the book is that the 'hegemonic despotism' thesis is somewhat overstated. While it may accurately describe the conditions for some kinds of workers in certain places within the capitalist system, it is far from an accurate representation of the lot of *all* workers. In particular, the thesis masks the many layers of geographic complexity that characterize the

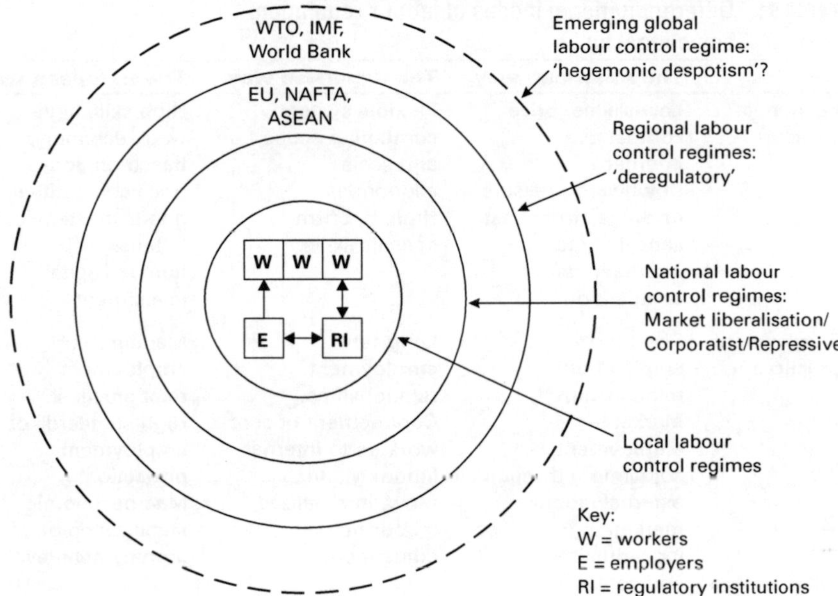

Figure 5.2 The 'scaling' of labour control regimes

global capitalist economy. Using the analytical framework developed in Chapters 2–4, we can consider this complexity at several spatial scales. 'Hegemonic despotism' is perhaps best understood as a dominant *tendency* in labour–business relations worldwide. Importantly, as Peck (1996) points out, even when the regime is not dominant within national states, neoliberal ideals – as promulgated by institutions such as the IMF and WTO – frequently shape the nature of relationships *between* them.

However, profound geographic variability is evident at four smaller and 'nested' spatial scales (see Figure 5.2 for a visual representation). First, *macro-regional* labour control regimes may show important variations. While it is true to say that most regional blocs (refer back to Box 4.6) are pursuing a deregulatory agenda, they differ in the extent to which they institutionalize labour rights and protection. For example, while the labour component of the North American Free Trade Agreement (NAFTA) is largely non-existent and has been described by one commentator as 'toothless' (Moody, 1997: 127), the European Union (EU) at least has a social charter that recommends certain basic rights and conditions for workers in signatory countries. Second, *national* labour systems with different institutional and

TABLE 5.1 Different national modes of labour regulation: some ideal types

	The American way	The Japanese way	The European way
Economic structures	Low-skilled, price competitive approach. Downward pressure on wages from cost sensitivity to international competition.	Flexible system, combining scope and scale economies. High, but firm-specific skills.	High-skill, high-wage economy, based on non-price competitive goods and services. Extensive human capital investment.
Employment practices	Short-term employment relationships. Market-led employment adjustment through external labour market. Individualized and competitive employment relations.	Long-term employment relationships. Confinement of core workers to internal labour markets. Slow, internalized labour market adjustment.	Medium-term employment relationships. High standards of employment protection. Macroeconomic stabilization of employment levels.
Economic and social norms	'Market' work ethic. Increasing working time combined with low productivity growth, and falling investments in human capital. High levels of labour market inequality.	'Corporate' work ethic. Acceptance of long working hours. Submission to corporate authority. Implicit socially encoded rules of managerial dominance.	'Social' work ethic. Declining working time. Low inequality. Universalized basic income levels.
Industrial relations	Market based. Individualistic ethos. De-unionized, human resource management. (HRM) approach.	Ethos of culturally and institutionally embedded trust, cooperation and compliance. Enterprise unionism.	State-articulated regulation of economic and political relations. Institutionalized rights of employment. Unions conferred with public status.

Source: abridged version of Peck (1996: table 8.1)

cultural histories may vary greatly in the extent of their 'despotic' tendencies. For example, Peck (1996) identifies three ideal-types of national system – 'American', 'Japanese/Asian' and 'European' and their chief characteristics (summarized in Table 5.1) against which different national contexts can be compared. While the American system adheres most closely to the notion of hegemonic despotism through its emphasis on market deregulation, the other two differ in significant ways that may ameliorate the position of workers somewhat. Importantly, both are more 'corporatist' in the sense that worker representation is institutionalized and state regulated. In the Japanese system worker organization is largely based upon firms, while in the European model it is centred on independent labour unions. Following Suárez (2001), by casting the net wider to include a range of newly industrializing economies (NIEs), we should add another kind of regime to Peck's typology, namely a 'repressive' type characterized by the aggressive imposition of constraints to achieve the control and subordination of labour.

Third, as described in Part One, there is considerable *place variation* in the character of labour–capital relations. The declining ability of unions to challenge employer practices has been far from even – and indeed many localities have never been unionized in the first place – with the result that some places remain more resistant or immune to the spatial restructuring and capital mobility that typify hegemonic despotism than others. Equally, the strength of ties between employers and local communities varies greatly geographically. Where firms have committed substantial sunk costs and enjoyed progressive workplace relations, the position of labour may seem far less bleak (see Chapter 6). In short, no two local labour control regimes are exactly alike, and the precise problems and possibilities facing workers vary accordingly. Fourth, there are of course important *intra-place* variations in the incidence of 'despotic' characteristics. In a global city such as New York or Los Angeles, garment workers locked into a harsh regime of global cost-competition may be working only a short distance from bankers or financial analysts working in a service industry that simply does not correspond in any way to the notion of a 'race to the bottom'. This leads to a final layer of complexity that cross cuts these scalar variations. The specific production systems of particular *firms* and *sectors* will determine the extent to which they can engage in aggressive spatial restructuring strategies. For every toy manufacturer that can switch production quickly between different subcontractors, there may be a car assembler tied spatially into complex supply chains, and many more service firms rooted in place by the need to be near customers or highly-skilled employees.

In sum, while there may be a *tendency* towards hegemonic despotism in global capitalism, driven by a powerful series of technological, regulatory and competitive dynamics, by mobilizing the conceptual framework from Part One we are immediately able to see that the extent and nature of the impact of the trend will be highly geographically – and therefore socially – variable.

MECHANISMS OF LABOUR CONTROL

Notwithstanding our critique of the hegemonic despotism thesis, it is undoubtedly true that workers in contemporary capitalism are faced with a broad range of labour control mechanisms that either threaten to erode the terms and conditions under which they work, or worse still, remove their opportunity to work at all. In this section we review the different mechanisms that can be used by both firms and states in this regard. Much could be written about these processes and their impacts on worker livelihoods around the globe. Here though, the aim is simply to illustrate the *range* of dynamics through which the position of workers *may* be diminished. Equally, of course, many of the processes detailed will have – theoretically at least – a more positive corollary or upside that *in certain contexts* can work to the advantage of workers (we shall see instances of this later in the book). For the purposes of clarity, we will divide our analysis into two parts, corresponding to the control strategies utilized by firms and states respectively. In reality, however, there may be differing degrees of collusion or conflict between the two groups as they seek to shape the contours of a particular labour control regime.

Firm strategies for labour control

Many of the most detrimental dynamics for workers come about as the result of firm restructuring and reorganization. Following Dicken (1998: 219–20), we can identify two broad categories of stimuli that may lead to firm restructuring – *external* and *internal*. External conditions may be either *negative* – as in the case of declining demand for a firm's products, increased competition, increasing input costs, labour resistance or militancy, and tightening government regulations – or *positive*, as with the enhanced opportunities for trade that may open up as regional blocs in which firms are located develop. Internal pressures may also lead to restructuring, and may relate either to the firm as a whole, or to a constituent element. Certain elements of production may be too costly, for instance, or sales figures may be below targeted levels. Interestingly, internally stimulated

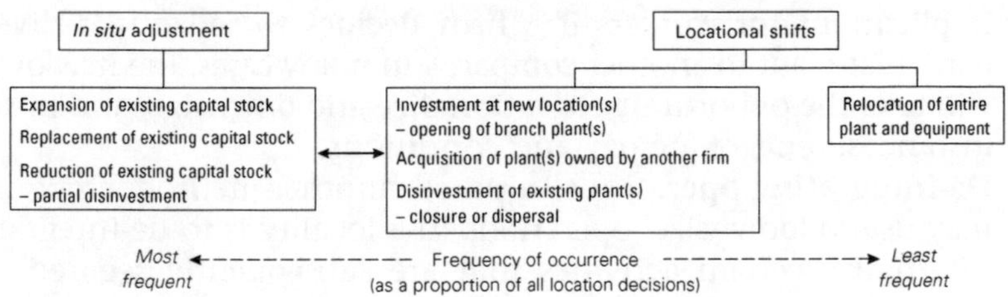

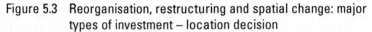

Figure 5.3 Reorganisation, restructuring and spatial change: major types of investment – location decision

Source: Dicken, 1998: figure 7.9

restructuring often seems to be associated with changes at the management level and consequent changes in a firm's strategic direction (Dicken, 1998). While the distinction between 'external' and 'internal' stimuli is useful analytically, in reality the two sets of factors are closely entangled and almost impossible to separate. The decision to close a particular plant, for example, may simultaneously reflect a combination of new strategic thinking *and* competitive conditions in that particular area of activity. Furthermore, individual firms may respond very differently to the same set of stimuli, which points to the idea that firms develop their own, unique decision-making cultures (Schoenberger, 1997).

There is a broad range of restructuring strategies that firms may pursue (Enderwick, 1989). Here we will classify them into two types: those associated with altering the *scale* of operations in places, and those that involve changing the *nature* of operations in places.

Changing the scale of operations 'in place'

- **Leaving place:** most obviously perhaps, firms may simply decide to relocate production to another locality. Except in the extremely rare occasions where the new site is a commutable distance away, relocation tends to result in the displacement of the existing workforce. Whether those workers will be able to find jobs on similar pay and conditions locally, or indeed at all, will depend on the nature and health of the local economy.
- **Rationalizing in place:** in reality, as Figure 5.3 indicates, and as was argued in Chapter 3, the decision to 'up and leave' a place completely is one very rarely taken by firms. Instead, firms are more likely to change the scale of operations in a particular place, or, in the case of multi-plant firms, alter the balance of activity *between* various locations. When firms chose to scale back operations in a place, workers are laid-off as before. There may also be

implications for workers if a firm decides to sell off or divest a particular plant to another company. In many cases, the new owners will take the opportunity to rationalize and downsize, and in some instances, renegotiate pay and conditions.

- **De-integrating operations in place:** another method which firms may use to 'downsize' operations in a locality is to de-integrate or externalize certain activities that are subsequently secured from suppliers or through subcontractors. This process presents two main problems to workers. Most profoundly, if tasks are externalized to a provider outside the locality, jobs may be lost. On the other hand, where the activity is still undertaken locally, workers may find themselves working for suppliers and subcontractors for worse pay and conditions, due in large part to increased levels of inter-firm competition. The result is a growing peripheral workforce characterized by casualization, and in extreme cases by new retrograde forms of labour such as homeworking and informal work.

Changing the nature of operations 'in place'

- **Technological change:** as we saw in Box 4.4, recent technological developments have had profound impacts on the nature of production systems. The *in situ* implementation of new production technologies can impact on workers in two key ways. On the one hand, they may simply replace workers and result in job losses. On the other hand, they may alter the labour process for the workers involved. While some commentators put a positive spin on this process, suggesting that mundane tasks are replaced and skilled positions are created, critics suggest that new technologies have increased the potential for worker surveillance and monitoring. For example, machines in a car assembly plant may be able to record the productivity of individual operators.
- **Intensification of working practices:** closely related to the implementation of new technologies, firms may seek to intensify the labour process of their employees in other ways. Examples include moving to new and extended shift systems, increased use of overtime, the lowering or ending of barriers between different jobs (known as 'multi-skilling', which may increase the insecurity of workers performing tasks now able to be done by others), and the use of a range of more 'flexible' employment contracts such as part-time, short-term and temporary working.
- **De-unionization:** firms may employ strategies of varying severity to reduce the role of unions: (1) they may simply ban unions completely in the workplace; (2) they may establish one-union rather than multi-union workplaces; (3) they may insist on workplace

bargaining and dispute resolution as opposed to collective bargaining at larger spatial scales; and (4) they may establish bipartisan worker-management groups councils within the workplace that seek to reduce the need for union representation.

- **Concessionary bargaining:** both intensification and de-unionization serve to weaken the position of workers when it comes to bargaining with employers. As a result firms are able to pursue *concessionary* bargaining strategies in which they seek to freeze and in some cases reduce wages. The *discursive* strategies introduced in Chapter 4 clearly come into play here as well: firms can help to enforce concessionary bargaining by emphasizing the perils of global cost competition and the threat of relocation during negotiations.
- **Selective recruitment and worker migration:** as recognized in the spatial divisions of labour thesis, firms often establish operations in places to access certain types of labour. Of particular concern are situations in which firms seek to tap pools of cheap and weakly organized female workers, then use the practices described above to hold down pay and conditions. Groups of migrant workers may be similarly exploited due to their need for work and relative lack of local social contacts. In some US cities day labour agencies shuttle workers from the poorest communities to perform low-paid work in the more affluent suburbs, charging them for the service (Peck and Theodore 2001).

Kim Moody uses the term 'lean production' to describe the nature of the production system in contexts where most or all of these eight categories of mechanism coincide. The deleterious impacts of this system for workers have received critical attention from a number of leading labour scholars and economists (see Box 5.1).

Box 5.1 Working in a 'lean' world

American labour scholar and activist Kim Moody (1997) provides a compelling critique of the impacts of 'lean production' upon workforces around the world. The term was first coined by Womack et al. (1990) to describe the nature of new production methods emerging from postwar Japan. The essence of lean production is that it combines economies of scale with economies of scope derived from various practices of 'flexibility'. Three types of flexibility are encapsulated in lean production. First, *functional* flexibility concerns increased levels of job rotation and multiskilling. Second, *numerical* flexibility reflects a growing use of 'contracting out' and part-time and casual labour. Third, *temporal* flexibility describes the shift to new forms of shift patterns (such as 24-hour working) and overtime. These changing working patterns are often combined with the very latest productive technologies. Moody's argument is that flexibility is most often implemented to facilitate cost reduction, not product and process innovation, as proponents would claim. Despite seemingly positive descriptors such as 'quality management' and 'team-working', these changes are not about empowerment, but about the micro-management of workers' time and activities, a feature Moody describes as 'management-by-stress' (p. 87). New forms of groupings designed to increase worker input to the production process such as teams and job circles can actually serve to bypass unions, or avoid them altogether. In some situations workers and their representatives are forced to develop proposals to reduce costs and improve quality in competition with other plants. The usual outcome of lean production is a combination of job losses, and increased work and time pressure on the primary workforce. For other critical accounts of lean production, see Gordon (1996) and Harrison (1997).

State strategies for labour control

These varied processes of corporate restructuring may be supported – or not, as the case may be – by state and quasi-state bodies at a range of scales. In particular, as we saw in Chapter 2, the national state remains of critical importance in any analysis of global capitalism. Six interlinked categories of mechanisms can be identified, many of which are associated with the re-scaling and 'hollowing-out' of the state described in Chapter 4:

- **'External' macro-economic deregulation:** by removing or lowering regulatory barriers to trade, foreign direct investment and the movement of people, technologies and capital, states (and quasi-state bodies such as the WTO and IMF) are creating conditions that allow productive capital to be switched between places at the international scale. Where macro-economic deregulation is very rapid the consequences for workers can be extremely grim (see Box 5.2 on the problems of economic restructuring in the post-socialist economies of east and central Europe).
- **Re-scaling of economic competition:** the 'downwards' re-scaling of certain state functions to the subnational level has increased inter-place competition for investment. In particular, the emphasis now placed on economic development and place marketing at the local level means that places within the same country are often competing in a zero-sum game for inward investment.

Box 5.2 Restructuring in transition economies

The problems of deindustrialization are somewhat amplified in the context of the so-called 'transition economies' of Eastern and Central Europe. Since 1989, countries in this region have been undergoing painful adjustments as they move from state socialism and centrally planned economies towards a capitalist market economy. For Bradshaw (1996), there are four inter-related dimensions to this transition. First, *economic liberalization* (or deregulation) has seen the removal of government restrictions on many areas of economic activity, but most importantly, in terms of relaxing fixed pricing systems. Second, attempts to establish *macro-economic stabilization* by controlling inflation and government finances are ongoing. Third, widespread *privatization* has seen the transfer of formally state owned enterprises – which in many cases constituted the entire economy – into the private sector. Fourth, an ongoing *internationalization* of the economy has seen the state retreat from being an active participant in trade to simply a regulator. As a result, levels of international trade and inward foreign direct investment have increased dramatically. While the experience of transition has been uneven, generally manufacturing industry in these countries has fared badly, with sharp downturns in production and employment levels. Smith and Pickles (1998: 7) describe the consequences of 'transformational recession' for the workers and citizens of these states in stark terms:

> The experience of 'transition' ... has been ...one of economic collapse, an onslaught on labour, and social and political disorientation ... The result has been a profound increase in poverty and inequality ... real wages have dropped dramatically and wage differentiation increased. Alongside increased inequality, homelessness has risen, health levels have declined and other social problems associated with polarization have emerged.

- **'Internal' labour market de-regulation:** states may choose to alter the nature of their regulatory intervention in labour markets. In particular, they may seek to enhance the flexibility of labour markets through removing constraints to labour mobility between jobs and market-determined wages. For example, basic protections such as a legislated minimum wage may come under threat.
- **Union re-regulation:** states may seek to deliver labour market flexibility by using toughened legislation to constrain the activities of unions and collective bargaining. Policies may range from an outright ban on unionization, to those that remove the automatic right to union representation, national bargaining, or the ability to call a strike.

- **Rolling back of welfare support:** another component of a move to labour market flexibility may be to change the nature of support available to those without work. Some industrialized countries such as the United States and the UK are seeing welfare systems replaced by so-called *workfare* systems in which the unemployed are enrolled into a variety of temporary training schemes and low-end jobs in order to 'earn' their benefit payments and hopefully secure long-term employment. According to critics, this shift simply provides an army of 'low-end' workers to be exploited by employers (Peck, 2000).
- **Privatization and downsizing the public sector:** finally, we must not forget the fact that the state is still a massive employer in most countries. Three dynamics may be significant here for labour control. First, any moves to reduce the public sector may result in redundancies, and relatedly, where services such as education, health and housing are concerned, have negative impacts in the reproductive sphere. Second, the 'contracting-out' of certain activities to external providers brings into play the same range of dynamics considered above in the context of firms. Third, the wholesale privatization of previously government-owned industries can expose workers to the forces of inter-firm competition.

To summarize this subsection briefly, we have reviewed the broad range of corporate and state strategies that can have detrimental effects on workers in certain contexts. Importantly, they range well beyond the mechanisms of capital mobility and inter-place competition that are given such prominence in accounts of 'hegemonic despotism'.

CASE STUDIES OF GEOGRAPHIC COMPLEXITY

We now have an appreciation of the range of mechanisms that firms and states may employ to extract concessions from workers in contemporary capitalism. There is no way we can undertake to 'map out' the impact of these processes on various types of workers in the different parts of the world in an anywhere near comprehensive way here. In the remainder of the chapter our approach is therefore to present three case studies in order to make three *geographical* arguments about the impacts of these labour control mechanisms. Taken together, the three studies augment and provide more depth to the central argument that the position of workers varies significantly from place to place. First, we will use a case study of industrial restructuring in Northeast England to illustrate how many of these different mechanisms, in reality, *operate in conjunction* in particular places. Second, we will examine the nature of service sector labour markets in Southeast England to reveal how labour control mechanisms can *impact differentially* within a particular locality. Third, using evidence from three newly industrializing localities in Southeast Asia, we will illustrate how the exact package of labour control mechanisms used *varies* from place to place, even when those places occupy broadly similar positions within international spatial divisions of labour. In particular, the aim of these case studies is to emphasize the *problems* that global capitalism poses to workers in these different contexts. We consider the various *possibilities* that workers have in the remaining chapters of the book.

Living with 'lean and mean' – industrial restructuring in Northeast England

The first case we will consider is industrial restructuring in Northeast England, a region that shares important characteristics with many other 'old industrial spaces' in the de-industrialising zones or 'rust-belts' of Western Europe and North America. These are areas where capitalist production expanded rapidly in the nineteenth century around 'old' industries such as coal mining, chemicals, iron and steel and metal processing (for example, shipbuilding, railways, engineering, etc.). The defining characteristic here is employment decline in the manufacturing sector of large urban metropolitan areas, closely accompanied by rises in registered unemployment rates. While the reasons for the reduction in manufacturing employment are many, it is usually attributed to varying combinations of global competition and productivity increases (see Box 5.3). Many regions have suffered declines in successive waves of manufacturing investment, leading to entrenched long-term unemployment in portions of the population and high levels of income inequality. However, to portray the employment situation in these areas as simply bifurcated between those in full-time waged employment and the long-term unemployed is a dramatic over-simplification. A wide range of labour control mechanisms have been implemented, resulting in complex and highly segmented labour markets characterized not only by the number of people without jobs, but also great variability in the character and quality of employment opportunities.

Box 5.3 Cheap labour or technological change?

This is a pertinent moment to revisit the 'myth' of cheap labour (see Chapter 1) which holds that manufacturing job loss in industrialized countries is largely due to competition from cheap foreign labour and subsequent increases in imports. However, this argument deflects attention from impacts of technological and organizational restructuring in those countries. In reality, many job losses have been due to changing competitive positions among firms in industrialized nations, the reorganization of work (e.g. adopting processes of multi-skilling), and most importantly, productivity (output per worker) increases due to the implementation of new technologies. Changes in levels of domestic demand and exports will also contribute to the overall level of employment change. At an aggregate, economy-wide level, economists estimate that only between 10–20 per cent of manufacturing job loss in industrialized countries has been due to 'trade effects', with the remainder due to 'domestic' sources (Hirst and Thompson, 1999). However, disentangling these different employment effects is difficult, as they are not entirely independent of one another. For instance, some of the increase in productivity due to technological change will have been stimulated by the pressures of overseas competition, in particular from lower cost producers. Whatever the exact level of trade effects, the impacts will, however, be sectorally and socially uneven. For example, studies considering the impact of the North American Free Trade Agreement on Canada's labour markets suggest that while the overall impacts on employment will be small, the job losses will be confined to particular routinized manufacturing sectors where workers are in direct competition with lower Mexican wages. The effect may be to further intensify Canadian wage polarization between knowledge workers, and lower or unskilled manufacturing workers who are already on low wages (Betcherman, 1996).

We can explore this complexity by looking at the economy of Northeast England (see Figure 5.4). The economy of the region, historically based upon the 'old' industries of coal mining, iron and steel and shipbuilding, has seen profound changes in composition over the last 30 years. Over the period 1975–95 there was a net loss of 179,000 jobs – equivalent to 14 per cent of the 1975 total – that resulted in the unemployment rate stabilizing at around 10 per cent for much of the 1990s. Moreover, taking into account those who have 'fallen out' of the labour market due to long term illness or permanent sickness meant that almost 33 per cent of the region's working age population were inactive in 1997. In sectoral terms, from 1975–95 some 57,000 or 73 per cent of jobs in the energy sector and 31,000 or 34 per cent of construction jobs were lost. Despite considerable investment from

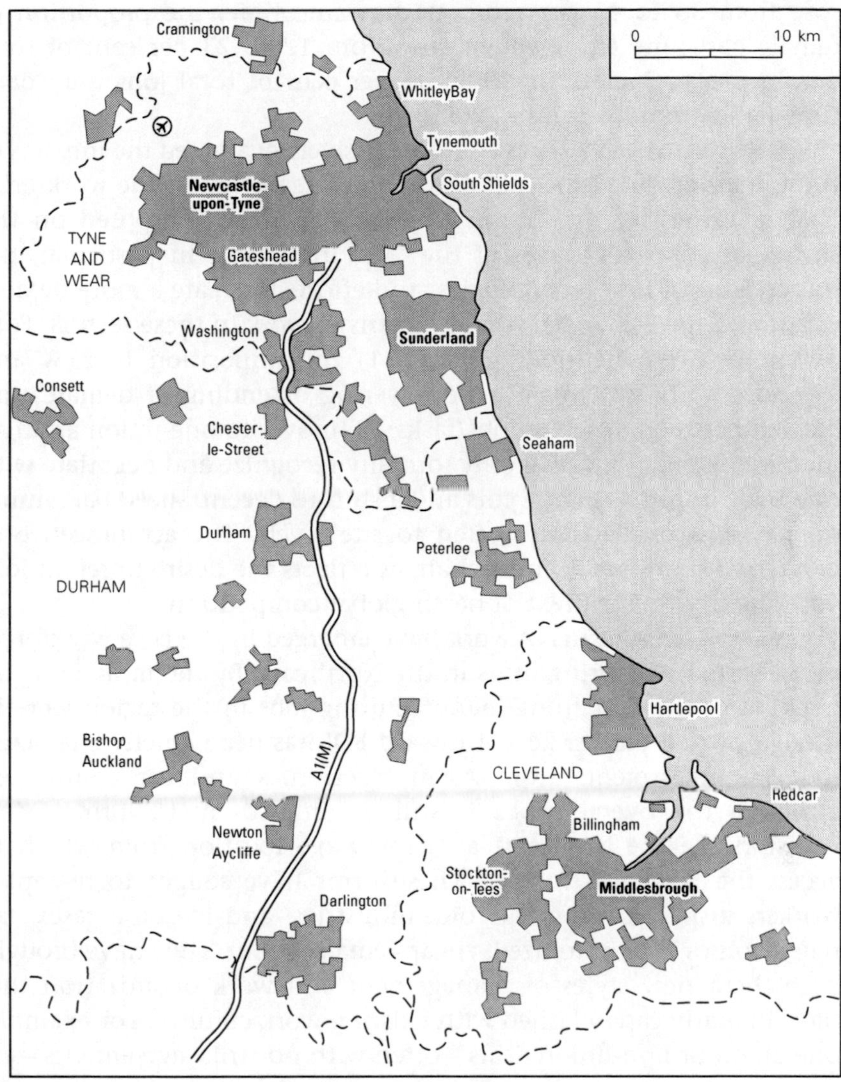

Figure 5.4 North East England: selected locations

abroad creating employment in 'new' industries such as car manufacturing and electronics, overall 202,000 or 45 per cent of manufacturing jobs disappeared. This decimation of the traditional employment base was only partially offset by the addition of 117,000 service sector jobs to the regional economy. Other dynamics were cross-cutting these trends. Over the period 1971–97, male employment fell from 64 per cent of the regional employment to 52 per cent, as female employment rose from 36 to 48 per cent. At the same time, the proportion of female part-time employment rose from 12 to 23 per cent of total employment. Overall, by 1997, 30 per cent of total jobs were part-time (all data from Hudson, 2000).

Ray Hudson (1989) has comprehensively evaluated the impacts of these restructuring processes on different segments of the workforce. First, a variety of new forms of work have been imposed on the shrinking core workforces of the 'old' industries. In particular, the impositions of new technologies, and efforts to create a more flexible workforce, have changed what it means to work in these sectors. Four dynamics need mentioning here: (1) the imposition of new and extended shift systems; (2) the lowering or ending of demarcation barriers between different jobs/tasks; (3) moves to one-union arrangements wherein a firm will only formally recognize and negotiate with one trade union; (4) wage cuts and a shift to decentralized bargaining on pay and productivity (often to site level). The acquiescence of workers and unions to these changes reflects the desire to retain jobs and capacity in a context of harsh global competition.

Second, new forms of work have emerged in the core workforces of 'new' manufacturing firms in the Northeast. By the mid-1990s, 33 per cent of the remaining manufacturing jobs in the region were in foreign owned plants. Recent inward FDI has been chiefly characterized by investments from Asian electronics and car companies, although the overall stock is still dominated by US firms. Most investors see the Northeast as a low-wage location from which to access the European market. These firms have sought to re-deploy workers displaced from the 'old' industries, and in some cases, tap new pools of non-unionized, cheap female labour. They have brought with them new styles of management and work organization that have in many cases clashed with existing work cultures. For example, one union or non-union deals – often with no-strike agreements – are prevalent in these sectors. There are also debates surrounding the quality of jobs on offer. At the Nissan factory in Washington, Wearside, the company presents an image of an enriching work experience within a lean production system built around notions of flexibility, quality and teamwork. According to critics, by contrast, the labour process is intense and tightly controlled, and operating in the context of low levels of unionization (see Box 5.1 again). While job creation from inward investment has at least soaked-up a proportion of the jobs lost in indigenous manufacturing, these foreign plants are themselves far from safe from rationalization and even closure. A well-publicized example was the closure of the Siemens wafer fabrication plant in Wallsend, North Tyneside in 1998, due to competitive pressures in the semiconductor industry. The plant, which attracted some £60 million in local and central government financial assistance, had only opened in 1996. The result was that 1,200 jobs were lost less than two years after they had been created! In turn, the plant was bought by US company Atmel in 2000, this time with the aid of £27.8 million of UK central government assistance, illustrating the cyclical nature of the inward investment process (Dawley and Pike, 2001).

Third, a growing peripheral or 'tertiary' workforce has emerged as businesses have responded to competitive pressures by 'externalizing' production to networks of subcontractors, a process occurring in both 'old' and 'new' manufacturing firms alike. For this peripheral workforce, inter-firm competition puts downward pressure on workers' wages and conditions. Indeed, at the margins of the system, growing numbers of workers are working 'off the cards' (working informally for cash payment), or being placed through 'temp' agencies on short-term contracts.

Fourth, the growth in part-time workers has been fuelled not only by the changes in the manufacturing working practices, but also through considerable employment growth in private sector services, and in particular business services, tourism, leisure, retailing, and back offices/call centres. In terms of the latter, developments have seen several thousand jobs created in what amounts to a new form of 'branch plant' economy as national service providers seek to tap pools of relatively cheap female labour for routinized tasks. Overall, the new employment in services is characterized by poorly paid non-unionized jobs, mostly offered on a part-time or casual basis.

These changes to the economy of Northeast England over the past three decades have been profoundly shaped by dynamics of change at larger spatial scales, and in particular, the restructuring of the state in the UK. There are three inter-related dimensions that had particular implications for old industrial spaces such as the Northeast (Hudson, 2000). First, the ascendance of neoliberalism associated with the Conservative government installed in 1979 had significant impacts. At a time of severe global recession in the early 1980s, policies designed to lower inflation and maintain a high sterling exchange rate exacerbated the competitive pressures on UK manufacturing leading to widespread job losses. At the same time, the deregulation of outward foreign direct investment precipitated a massive outflow of manufacturing investment in the 1980s to cheaper offshore locations. Second, the privatization and rationalization of the previously government-owned coal and steel industries opened these sectors to the full force of international competition, accelerating the long-term decline of these sectors. Third, the Conservative government's desires to pare down the public service sector led to job losses and worsening terms and conditions in what traditionally was the most important form of service sector employment in regions such as the Northeast.

In sum, the example of the Northeast's economy serves to show how workers in some places are seemingly faced by the full battery of labour control mechanisms described earlier in the chapter, operating in conjunction. In responding to the pressures of international competition, both 'old' and 'new' firms alike have consciously altered both the scale and nature of their operations, with a range of retrograde impacts on workers, processes that have been aided and abetted by the neoliberal nature of the national policy regime in the UK.

Two sides to growth – service industry work in Southeast England

Our second case study explores the nature of service sector employment in Southeast England. The region is sometimes characterized as a 'new industrial space', a term used in economic geography literature to delimit the localities within industrialized countries that have seen a resurgence of economic growth over the last two decades. New industrial spaces have typically been associated with three groups of 'propulsive' industries: namely, the manufacture of high quality design-intensive goods such as jewellery and leather items, high technology manufacturing (for example, a semiconductor or computer firm), and, as in the case of Southeast England, with concentrations of financial and business service employment. The term financial services refers to activities such as banking and insurance, while business services are a more diverse group including the older 'professions' such as law, accountancy, architecture, surveying and engineering consultancy, as well as newer activities such as staff recruitment agencies, market research, computer services, security and cleaning.

In the initial theoretical formulation (Scott, 1988), new industrial spaces were seen to have five key characteristics: a strong spatial agglomeration of activity in particular sectors, a strong social division of labour, a proliferation of small to medium sized firms, dense networks of externalized transactions between firms, and the development of a relatively self-contained labour market. While some of these suppositions have come under subsequent challenge – in particular, the importance of large firms, and extra-local corporate and market

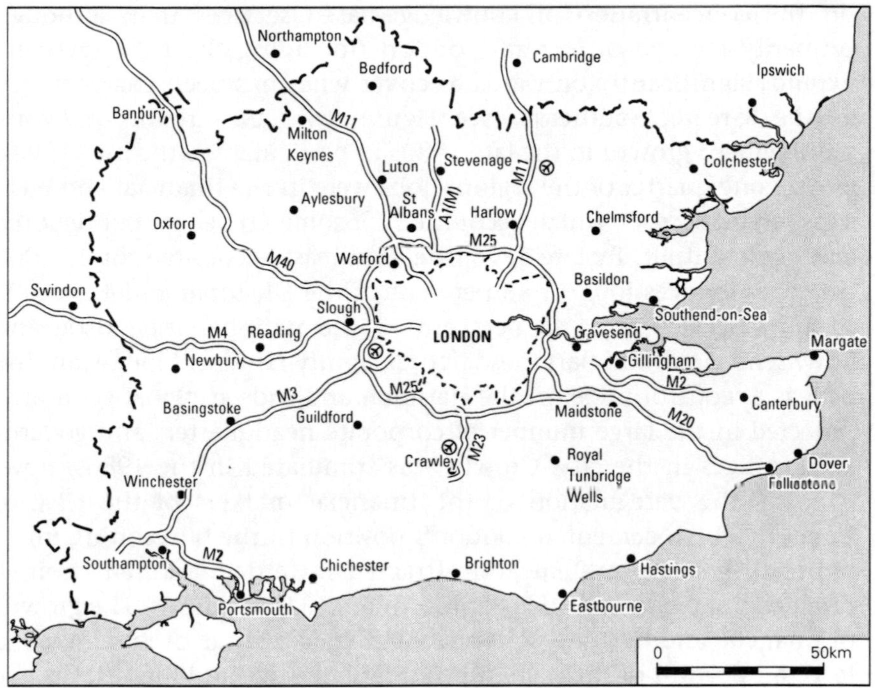

Figure 5.5 England's South East standard region

connections in many of these spaces is now well documented – the thesis has provoked debate about the nature of local labour control regimes (LLCRs) in these 'hotspots' of employment growth. It appears that, in many cases, through the widespread adoption of 'flexible' labour practices such as individual bargaining, multi-skilling and the use of part-time or temporary staff, LLCRs characterized by high levels of worker turnover, low rates of unionization, and dependence on large pools of relatively marginalized support workers are developing. In short, these labour regimes, while lauded by some for their progressive elements, can be just as deeply segmented markets as those in other contexts (Crang and Martin, 1991). The labour force may also be strongly gendered, with the managerial-technical class being very male dominated, and many of the support tasks being highly feminized (Massey, 1995b).

Southeast England provides a useful window on the downsides for labour within these growth zones. The region currently contains the leading agglomerations of both high technology industry, and financial and business services, in the UK economy. Here we will focus on the concentration of knowledge-based services that, although primarily focused on central London (including the City), actually extends significantly outwards to cover what has been characterized as the 'Greater Southeast' (see Figure 5.5). As a result of strong employment growth in the late 1980s in particular, by the mid-1990s, almost one quarter of the region's jobs were in the financial and business service sectors, which accounted for some 40 per cent of regional economic output. By late 1996, the Southeast accounted for 468,000 financial services jobs, or 46 per cent of the UK total, and 1,267,000 business service jobs, 45 per cent of the national total (Coe and Townsend, 1998). In part these figures signify London's long-standing role as a 'control node' in the national and indeed global economy, reflected in the large number of corporate headquarters and government offices in the city. Growth was stimulated in the 1980s, however, by the deregulation of the financial markets of the City of London, which cemented London's position in the burgeoning international financial system and attracted inward investment from a range of US, Japanese and European financial institutions. The growth of financial and business services has created a range of jobs. While a large proportion of them are full-time, high wage and high status, we must be aware that these firms also employ substantial numbers of receptionists, secretaries, typists and clerks on considerably lower wages. By and large though, these sectors constitute what the economic geographers John Allen and Nick Henry (1995) characterize as the 'upside of growth' in the Southeast.

An important component of the 'downside' is the associated growth of employment in low pay and low status labour intensive services such as cleaning, catering and security (Allen and Henry, 1995, 1997). Throughout the 1980s and early 1990s, these sectors experienced strong and sustained employment growth: by the mid-1990s, they contributed some 200,000 workers to the Southeast's economy. The region accounted for between one third and one half of total national employment in each of the three sectors, suggesting close links to the 'upside' of growth described previously. In reality, a good proportion of this employment growth is accounted for by the externalization of previously in-house jobs to subcontractors from a variety of sectors including the public sector. The shift to *contract working* is hugely significant, however, as we shall see. The nature of the contract service jobs can be broadly characterized as follows: in cleaning, the majority are for women working part-time; catering is dominated by women working a variety of hours but often above part-time levels; and in security the workforce is dominated by men employed on a full-time shift work basis. An ethnic division of labour also crosscuts these gender differentials, with, for example, Afro-Caribbean and white workers dominating in security work. Importantly, what is being described here is a different dynamic to the processes of casualization (moves to short-term and temporary contracts) that are prevalent in some forms of low-end service work. By contrast, work in these sectors is regular and formal. However, it is also extremely 'precarious' in nature due to its contract basis (Allen and Henry, 1997). Contracts with clients' firms are usually of just one to three years' duration, and workers have no legalized right to re-employment after that period. In reality, most workers are either shifted to another client site by their contractor, or stay on the same site under new management, when contracts change hands. However, frequent contract changes can lead to negative changes in the terms, conditions and tenure of employment due to the highly competitive nature of the tendering process. In this way, permanent insecurity is a fact of life for these workers.

The firms employing these workers have steadily grown in size through the general expansion in the subcontracting of these services, and processes of acquisition. By the early 1990s each of the sectors was dominated by a handful of lead firms: in cleaning, the top five firms accounted for one-third of the UK market, in security three firms held 60 per cent of the market, and in catering the top three firms had secured 80 per cent of the market. Many of these firms had become large, transnational corporations, and some were foreign-owned. To give one example, one of the top five cleaning firms was a Danish multinational with 120,000 employees worldwide of which 12,000 were employed in the UK. These peculiarly 'hollow' multinational firms have large workforces that are distributed in small groups – say four or five cleaners, two or three security guards, per workplace – across many client sites. One catering firm, for example, had 22,000 workers spread across 3,300 sites in the UK. This process of splitting of the workforce into small 'fragments' at multiple workplaces reduces the potential for organizing and resisting employment conditions, despite the fact that their labour is absolutely central to generating profits for the contract service firms. Labour control is enacted at the workplace level through site or unit managers whose main task is to ensure the service is provided effectively on a day-to-day basis. Contact with regional or head office, or fellow workers on other client sites, is highly circumscribed.

There are three other dimensions to the social and political marginalization of these workers. First, growth in contract services has been shaped and enabled by the same neo-liberal tendencies in state policies described in the case of Northeast England. For Peck and Tickell (1992), national deregulatory processes have produced a dysfunctional regional economy characterized by polarization and 'overheating'. In particular, the deregulation of labour markets in the 1980s removed legislative barriers to the emergence of these low wage and poorly unionized forms of work. Second, the social construction of these tasks as 'unskilled' – and therefore unworthy of reward by promotion or wage increases – by employers, client companies, and indeed society at large reinforces their marginalization. This is linked to the relative 'invisibility' of these tasks, particularly in the cases of cleaning and security work, which predominantly take place when client premises are almost empty. Third, many 'low-end' workers may have difficulty in affording a suitable place to live in the extremely tight and expensive housing market of Southeast England (Allen et al., 1998).

In sum, the experience of working in the flexible labour market of a new industrial space will vary greatly between different groups of workers within it, a situation described well by Martin (2001: 467):

> For service professionals, increased flexibility is likely to mean self-determined 'flexi-time', split home-office working arrangements, high salaries, and generous social and non-wage benefits. For part-time cleaners, increased flexibility invariably means working unsocial shifts and hours, short-term contracts, loss of employment rights and entitlements, and depressed pay and conditions.

Holding down the global – labour control practices in Southeast Asia

For our third and final case study we switch to the newly industrialized economies (NIEs) of Southeast Asia. The widespread overseas relocation of labour intensive manufacturing production from the United States and Western Europe first became evident in the 1950s and was crudely characterized by the NIDL thesis (refer back to Box 4.3). Over the period 1953–94, the share of world manufacturing accounted for by 'developing' countries increased from 5 to 20 per cent (Dicken, 1998). In reality, the vast majority of this increased manufacturing capacity is accounted for by a relatively small number of developing states, the so-called newly industrializing economies. While the four Asian tiger economies of South Korea, Taiwan, Singapore and Hong Kong, along with Brazil and Mexico, grew to prominence first in the 1960s and 1970s, by the early 1990s they had

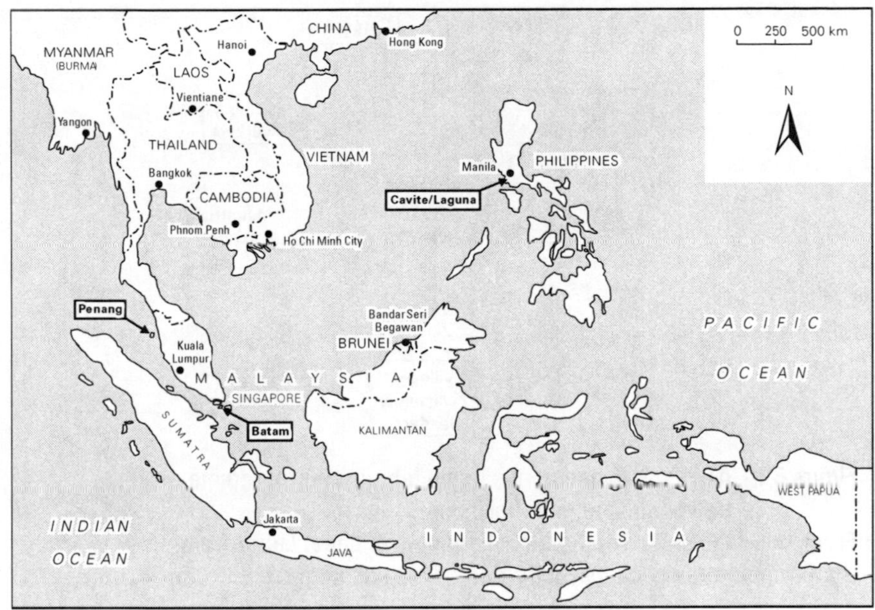

Figure 5.6 Case study sites in south east Asia

Source: Kelly, 2002: figure 1

been joined by a 'second wave' of Asian NIEs including Malaysia, Thailand, Indonesia, the Philippines, China and India offering lower labour costs. Of particular interest here are the labour control mechanisms used in the export processing zones and industrial estates within these new NIEs.

The key point we wish to make is that the package of mechanisms employed varies significantly between different production sites in the NIEs. The argument is developed in two stages, using the work of Canada-based economic geographer Philip Kelly. First, we consider in some detail how labour control is enacted in the provinces of Cavite and Laguna in the Philippines (Kelly, 2001). Second, we use Kelly's work to contrast the practices operationalized in Cavite and Laguna with those used in Penang, Malaysia, and Batam, Indonesia (Kelly, 2002). Taken together, the three localities share several important characteristics: the predominance of female employment in light manufacturing and assembly in the electronics and garment sectors; growth driven by foreign direct investment from North America, Europe and 'first wave' NIEs; heavy state involvement in the provision of locational and export incentives; and extremely low levels of unionization.

Figure 5.7 Actors and institutions in the labour control regime in Cavite and Laguna, Philippines
From Kelly, P.F. 2001. 'The Political Economy of Local Labor Control in the Philippines', *Economic Geography*, 77: 1–22, figure 2.

The provinces of Cavite and Laguna to the south of Manila (see Figure 5.6) have seen an intense period of foreign-investment fuelled industrialization in the 1990s that has added some 250,000 waged jobs to the economy. The transition from rice-growing hinterland to industrial zone has been accompanied by the emergence of a range of actors and institutions – operating at a variety of scales – geared towards the education, recruitment, training, discipline and reproduction of the workforce (Figure 5.7). Kelly's (2001: 7) analysis shows how the interaction between these various actors serves to 'suppress effective representation for labour and to assure corporate investors of a non-confrontational workforce'. Interestingly, and somewhat in contrast to localities in industrialized economies, many of the strategies used by the various actors are *informal* in their nature. While the Philippines actually has a relatively amenable set of national labour laws that allow union formation and activity, a largely union-free local zone of labour control is achieved through the inter-personal networks between a variety of actors.

Here we will briefly consider the role of each participant in the labour control regime in turn (Kelly, 2001).

- **National agencies:** two government agencies, the Board of Investments (BOI) and Philippine Economic Zones Authority (PEZA) are involved in the attraction of inward FDI and are the point of contact for foreign firms with national government.

While the former offers incentives to 'pioneer' and exporting firms, the latter manages the four state-owned export-processing zones in the Philippines, and regulates privately owned industrial estates. These agencies are important in three ways. First, they 'sell' the characteristics of the Filipino workforce – such as 'trainability' – to investors. Second, they intervene when labour disputes arise, and third, they can act as an intermediary between foreign firms and local or provincial government officials.

- **Corporate investors:** individual companies use a wide range of techniques to control labour. First, they may establish bipartite Labour-Management Councils to keep disputes contained within the firm and thereby pre-empt or nullify attempts at union formation at the company or inter-firm level. Second, firms may seek to maintain high rates of staff turnover to diffuse discontent and keep wages down. Third, firms may use inter-firm networks, in particular between human resource managers, to share and disseminate information on emerging labour issues. Fourth, they may extend their involvement into the reproductive sphere, for example by providing transport or health checks, in order to ensure compliance.
- **Factory workers:** certain kinds of workers tend to be selected by foreign firms in Cavite and Laguna. In particular, young, female workers with no previous industrial experience or union involvement are seen as being desirable. Selected workers are then carefully inculcated into the factory environment through orientation seminars and the like.
- **Industrial estate management companies:** the services provided by estate managers go way beyond the provision of infrastructure and factories, into the realm of labour control. There are several elements to this. First, known union organizers are largely excluded from estates by tight border security. Second, estate managers may assist firms by compiling databases of employable workers with a confirmed lack of union connections. Third, they may become directly involved in industrial relations, mediating between employer and employees. Fourth, managers liaise with local government officials and community leaders, thereby acting as a buffer between foreign firms and local politics.
- **Recruitment agencies:** most localities in Cavite and Laguna have a branch of the Philippine Employment Service Office (PESO), a government job-placement service provider. As well as providing 'screened' job applicants in much the same way as estate managers, these agencies can also help firms circumvent certain labour laws by offering temporary workers contracted to themselves rather than the firm itself.
- **Village/community leaders:** both firms and estate managers rely on local officials and leaders, in particular to certify whether

workers have ever had union involvement. Firm strategies that enter into parts of the reproductive sphere will also depend on liaison with community officials.

- **Municipal governments:** securing their backing of municipal officials such as mayors can be important for firms and estate managers. For example, the local police forces necessary to assist with estate security come under mayoral control. Equally, some mayors may become involved in accrediting potential workers and subduing potential unionizers.
- **Provincial governments:** similarly, provincial governors can be – depending on the personality involved – key figures in delivering a weakly unionized and strike-free workforce to investors. The provincial government may also act in tandem with PEZA to tackle labour disputes when they arise.
- **Labour organizations:** partly as a result of the dynamics described above, levels of unionization are very low in Cavite and Laguna, covering an estimated less than 10 per cent of the workforce. In 1998, of 210 firms in the Cavite Export Processing Zone, only seven were unionised and just 23 had Labour-Management Committees. Two further contextual factors restrict levels of unionization. First, many among the young workforce only see their employment as temporary, and second, high background levels of unemployment mean a large supply of willing – and often migrant – labour is available for employment.

Labour control in Cavite and Laguna, then, is enacted through a wide range of formal and informal mechanisms involving a variety of differently scaled actors. The result is the delimitation of 'enclaved' industrial estates in which a compliant, non-unionized workforce is delivered to foreign investors. How does this configuration compare with those found in Batam, Indonesia and Penang, Malaysia, however? In both cases, somewhat different configurations of strategies are apparent. In Penang, effective labour control largely resides in the use of four mechanisms. First, the use of domestic and foreign migrant labour provides a largely cooperative workforce that is both distanced from the 'distractions' of home and family, and numerically flexible. Second, the housing of migrant workers in hostels gives employers a measure of control over the reproductive sphere. Workers can be closely monitored, counselled where necessary, and provided with a range of religious, sporting and educational activities. Third, dispute resolution is contained at the firm level through the use of Japanese style 'enterprise unions' (see Table 5.1). Finally, there are a number of legislative and bureaucratic barriers to union formation at the national scale. Most starkly, union formation is subject to an outright ban in the electronics sector. The situation in Batam is closer to that in Penang rather than Cavite and Laguna. The workforce is almost entirely constituted by intranational migrants, a large proportion of which is housed in hostels, and firm-level labour-management councils are used to constrain disputes to that level. Batam is embedded in a national context in which union suppression has historically been achieved through militaristic rather than formal, legislative means.

We can make two points in summary. First, as we would expect, the kinds of mechanisms employed in these industrializing localities are considerably different from those employed in the industrialized contexts of the preceding case studies. Second, and most importantly, the precise combination of measures employed varies between different places undergoing similar processes of economic development. In the same way, we would expect the dynamics we described in the cases of Northeast and Southeast England to play out somewhat differently in other 'old' and 'new' industrial spaces respectively.

SUMMARY

The aims of this chapter were twofold. First, drawing on elements of the 'hegemonic despotism' script, we have illustrated the broad range of mechanisms through which firms and/or states constrain and downgrade the terms and conditions under which wage workers are employed. Second, three case studies have shown the geographic complexity that underlies the implementation of these labour control mechanisms. In particular, it is evident that there are tremendous intra- and inter-place variations in the configurations of mechanisms used, and which groups of workers they affect. This spatial variation also feeds into one of the arguments running through the book, namely that it is extremely difficult to make blanket judgements about what is 'good' or 'bad' for workers, as such judgements are context dependent, and can only be made in relation to the position of certain groups of workers in particular places. For instance, it is highly possible that managerial or skilled workers employed in a foreign-owned electronics factory in China are better off, in relative terms, than immigrant workers in a garment sweatshop in Los Angeles, or contract office cleaners in New York. While this spatial complexity is extremely interesting in its own right, it is also of profound practical importance as it suggests that how labour responds to control mechanisms will also be place dependent and therefore highly spatially variable. In the next chapter we consider the range of ways in which workers, employers and local institutions may act locally to protect or improve worker conditions, before considering extra-local strategies in Part Three.

six

Varieties of Localism

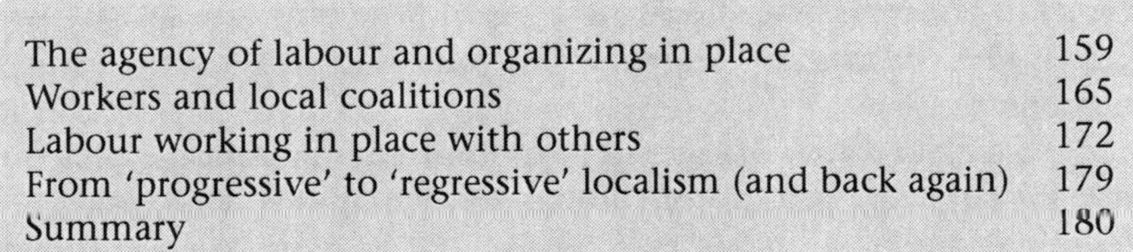

In the last chapter we explained how firms and pro-business regulatory bodies seek to set the conditions under which wage workers live and labour. And yet while workers might be experiencing declining living conditions in some places, they are not without *agency*. That is, workers, like firms and regulatory institutions, can act on their own or in alliance with others to secure better livelihoods. This capacity for agency tends to be downplayed in the influential discourse of globalization (the myths presented in Chapter 1), of which the idea of hegemonic despotism discussed in the last chapter is a part. In contrast, the emphasis in this chapter is on the ability of wage-workers to organize, to campaign, to keep jobs, or to attract new forms of business investment. Here we highlight the fact that 'workers may see their own self-reproduction as integrally tied to ensuring that the economic landscape is made in certain ways and not in others' (Herod 2001: 33).

If the ability of wage-workers to fight for better livelihoods is this chapter's key theme, then two others supplement it. First, we show that *local level agency* can be highly effective in furthering the cause of workers and their allies. This is what, in Chapter 4, we called *acting locally for local needs and wants*. The relative immobility of labourers is not, we argue, *necessarily* disempowering for them vis-à-vis employers and regulators. Second, we argue that it is wrong to see local level worker actions as *simply* politically 'regressive'. Many Left-wing commentators regard thinking and acting locally as perpetuating inter-place competition for jobs and investment – and therefore as undermining worker solidarity across space. While this judgement is a reasonable one, it is no simple matter weighing the gains and losses entailed. When workers in one place successfully retain or attract employment opportunities this cannot be categorically dismissed as a 'bad thing' because workers elsewhere may have lost out in the process. The real question is whether it is possible to weigh socio-economic benefits derived in one place against actual or potential benefits foregone elsewhere. Here again we come up against what constitutes 'justice' for workers. We return to this thorny issue in Chapter 9.

This chapter is organized into four main sections. The first revisits the idea, initially introduced in Chapter 3, that labour is a 'sentient spatial actor' (Herod, 1997a: 26). Workers are not animate objects or helpless pawns in the game of global capitalism, unable to alter their life conditions. Rather workers are able to exploit some of the 'wiggle room' we talked about in Part One. We distinguish here between 'pure' and 'synergistic' forms of worker agency. The former entails wage labourers acting alone; the latter involves them acting in concert with other local groups. After illustrating the first type of agency with an example, we then use the second section to introduce the different types of coalitions that can form in particular places. The third section then presents three examples of place-coalitions in action. In the first, workers ally themselves with local firms and regulatory bodies (in a so-called 'cross-class alliance'); in the second, they join forces with local civil society (in a so-called 'reciprocal' or 'community union alliance'); and in the third they ally themselves with local groups to operate *outside* of global capitalism (an 'opting-out' alliance). In this latter case, 'opting out' means rejecting capitalism as a whole. It is a radical and a rare form of local worker agency that is nonetheless a real option in certain places. What all the case studies discussed in the chapter share is that they are about *place-dependent* groups. That is, they involve workers and other local groups/institutions who have a very strong commitment to defending or promoting their locality. In the fourth section, the chapter argues that while each of our case studies shows that workers can extract concessions from firms and regulatory bodies, whether the effects of local level agency should be deemed 'progressive' or 'regressive' depends on who and where you are.

THE AGENCY OF LABOUR AND ORGANIZING IN PLACE

Workers have agency. They have the capacity to act, to change, to challenge and to resist. This may seem a surprising claim in light of the sorry stories recounted in Chapter 5. But these stories are by no means representative of the situation worldwide. In many places, workers have the capacity to decisively alter their circumstances. Put simply, wage labourers can be 'active participants' (Herod, 2001: 16) in the creation of the socio-geographical situations in which they find themselves. But what do we mean by the 'agency of labour'? The concept of agency refers to the capacity possessed by individuals and groups to act for their own benefit or for the well-being of others (see Box 6.1). To say that people have agency may seem so obvious as to be trite. After all, most readers of this book would have little difficulty listing countless examples of their freedom to think and act. However, it is important to note that the degree and kind of human agency is both *constrained* and socio-geographically *variable*. The constraints arise because all people think and act within sets of relationships that place definite limits on what is possible. We called these relationships 'structures' in Part One to highlight their relatively intransigent character. The sociologist Anthony Giddens (1979, 1984) has done most to draw our attention to how individual and collective agents act within wider societal structures – such as class, gender, and ethnic relations – and, through their actions, reproduce these structures. This is what he called a process of *structuration*, which Marx, in effect, described long ago when he famously said that 'people make history but not under the conditions of their own making'. An example relates to how some students end up at university while others do not. An analysis that privileges agency would look at the characteristics of the individual. Is the student bright? Does she work hard? It would conclude that those students who work hardest and that are the brightest are most likely to get to university. An analysis that privileges structure would look at the wider socio-economic context in which the student was educated. Did she come from a high-income family? Did her parents go to university? Did a large number of her school friends go to on to university? It would conclude that certain students are more likely than others to go to university, not just because of their individual traits, but also because of wider economic and social factors. A structurationist approach would bring together the two approaches, arguing that students go to university *both* because they are bright, *and* because in some way this brightness

Box 6.1 Human agency: the basics

In the last two decades debates over human 'agency' have been well rehearsed across the social sciences, from sociology and psychology to economics and geography. At its simplest, agency refers to the ability to act: that is, the capacity that individuals and groups have to make choices or decisions that, in turn, structure or shape their own lives. Agency is more, though, than people's intentions. It refers to the 'capabilities people have of *doing* things' (Peet, 1998: 155, emphasis added), thus combining intentionality and practice. According to Giddens (1979), all agency is enabled and constrained by social 'structures' which exist over and above the individuals and groups they affect. These structures are economic, political, cultural and/or social. They vary in their geographical scale, with some affecting most people worldwide (for example, global capitalism), and others varying from place to place (for example, political systems). All social structures, according to Giddens, form the 'unacknowledged conditions' that make individual and group agency possible. They are, if you like, the 'rules' that govern thought and action – they circumscribe what seems 'normal' and 'abnormal', 'possible' and 'impossible'. These rules permit a degree of openness in how people on the ground respond to them. For example, one of the 'rules' of the capitalist economic system is that firms must innovate technically if they are to survive. But individual firms can react to this pressure to innovate in all manner of ways. More specifically, Giddens argues that some groups have more 'resources' than others with which to express their agency. Thus wealthy, well-connected people have more agency than poor, marginalized individuals. In turn, Giddens argues, the cumulative effect of small acts of agency in different places is to reproduce – or sometimes transform – the social structures that are a precondition for agency in the first place. Thus all agency has what Giddens calls the 'unintended consequence' of maintaining social structures. Only during periods of extraordinary discontent can acts of agency radically alter structures. This is why, for example, social revolutions are called 'revolutions' (like that in Russia in 1917). They try to create an entirely new framework for people's everyday thought and action.

stems from factors such as family income, the time available for parental support, the quality of pre-university education and so on. As Giddens would have it, structures condition agents and agents, collectively, affect structures (usually reproducing, them but now-and-then transforming them). In light of this, it is plain to see why the degree and kind of human agency is socio-geographically variable. Imagine a hypothetical scenario where two people with the same mental and physical characteristics are born into a poor Sudanese family and a rich Australian family respectively. Who is likely to have the greater degree of agency over their life course? Who is likely to suffer more from structural constraints? Though many structures embroil most people on the planet – notably capitalism – it remains the case that within these structures some have a greater capacity to improve their life chances than others.

Interesting as this may be, what does this mean for an analysis of wage-workers? First, it draws our attention away from cases like those explored in Chapter 5. We are encouraged to look, instead, at where and how some workers take control of their lives and their destinies. Second, a focus on worker agency leads us to inquire into the importance of *geographical strategies*. Recalling our discussion in Part One, these are strategies employed by workers and allied groups that actively use or alter the existing geographical organization of economic activity at a range of scales. They are part-and-parcel of workers' power repertoires. They can range from acting in place (the focus of this chapter) to uprooting and moving elsewhere (see Chapter 7) to organizing trans-locally (see Chapter 8). So in this, and the following two chapters, we begin to appreciate that even though wage workers have to labour within the capitalist system, the geography of production and regulation is not the sole prerogative of those who are economically and politically powerful. Labourers too play an active role – sometimes alongside, other times in direct opposition to, business and regulatory bodies – in the way the economy is organized geographically. If, as we argue, this role can be performed in place or across space then intuitively it also makes sense to consider how labourers play a role in the construction of geographical scale. As we shall see in this and the next two chapters, re-scaling their actions, whether upwards or downwards, wage-workers are able to manipulate scale as a strategy for improving their lot under global capitalism.

Our specific focus here, as mentioned above, is the local scale (also the focus of the previous chapter). Operating alone or with other local stakeholders, workers can undertake actions in place that can palpably improve their fortunes. Of course, workers can also exert their agency inside of the workplace. They can resist some of the demands made on them by employers – either directly through strike action or through more indirect actions, such as slowing or even sabotaging production, ignoring management directives or by simply behaving inappropriately, such as laughing in meetings. Our attention here though is the extent to which the circumstances under which *local alliances* occur are explicable with reference to the idea of *place dependence* introduced in Part One. Wage-workers, their families and the institutions that support physical and social reproduction are at the very least place-based – that is, 'fixed' in the landscape and not readily movable – and often place-bound – that is, absolutely confined to the locality in question. Though many firms and regulatory bodies are not place-dependent in these ways, we saw in Chapter 3 that many others in fact *are*. It is these *joint* commitments to place – at once physical, financial and emotional – that can lead wage workers, local firms, local regulatory bodies and local non-working groups to come together in concerted actions to defend or promote their locality. So place can complicate the simple two-class model outlined early in Chapter 2, a model that depicts workers and firms existing in a mutually tense relationship with regulatory bodies somehow mediating in-between. In ideal-typical terms, then, we have two 'types' of local scale worker agency: 'pure' agency (where workers act alone, as it were) and 'synergistic' agency (where workers forge alliances with other place-dependent actors).

We deal with local alliances or what are called 'place coalitions' in the rest of the chapter. For now, though, we want to offer an example of wage workers undertaking local action alone. The example concerns a United Auto Workers (UAW) dispute with General Motors in Flint, Michigan – a place that was the subject of maverick film-maker Michael Moore's (in)famous movie *Roger and Me*. This dispute – one of literally thousands we could have chosen – is symptomatic of successful examples of 'pure' local worker agency in two respects. First, it involved workers who identified with each other not just because they shared the same employer but also because they belonged to the same trade union. As we argued earlier in the book, trade unions can be a major force for workers when they have the requisite resources and the legal room to manoeuvre on behalf of their members. Second, the Flint dispute was very much a workplace-focused struggle, rarely 'spilling out' into the wider city of Flint in any direct way. In a sense, the workplace is the most obvious site – or scale – at which workers can 'hurt' their employers or regulatory bodies that support those employers. It involves a very *direct* attempt to disrupt commodity production, distribution or sale 'at source'.

Just-in-time manufacturing and making a virtue of being 'local': the case of Flint, Michigan

Flint, like other places in the so-called North American 'rustbelt' lost thousands of manufacturing jobs during the 1980s and 1990s. As employers closed plants or relocated elsewhere so local communities found they were facing increasing levels of hardship. Against this backdrop, retaining large employers in the locale increased in importance for local workers and their dependents. Employers, on the other hand, saw that with every plant closure their ability to demand more from workers increased. They began to change the methods of work organization, with the introduction of new, more 'flexible' systems. Overtime payments were reduced, the timing and length of work shifts altered dramatically, and new rules governing shop floor production were established. At the same, so-called Just-in-time (JIT) supply systems were introduced. Previously, large auto-manufacturers like GM had stockpiled component parts for vehicle assembly. When vehicle demand dropped or there was over production, such inventories were costly to maintain not only in terms of storage but in terms of having to purchase so many extra components ahead of time. By contrast, with JIT systems a company like GM only purchases the exact number and type of components it needs from suppliers. These systems thus save companies money by removing the cost and the 'cushion' of large inventories. It was in this context that two GM plants in Flint chose to strike. Unhappy both with recent efforts to change the organization of work at both workplaces and with the amount of investment GM had made in the plants, strike action was organized.

The first strike began on 5 June 1998, when 3,400 UAW members stopped working at a metal stamping plant in Flint. Six days later, on 11 June, 5,800 UAW members at Flint Delphi Automotive Systems did the same. Within two weeks 193,517 workers at 27 of GM's 29 North American plants had been told to go home as there was no work for them to do, and 117 component supplier plants owned by subsidiaries of GM had either closed or reduced production. In addition to slowing GM's North American production, the firm's plants in Mexico and Singapore were affected, as the strike action disrupted the firm's highly inter-connected production system (see Figure 6.1).

Defining their geographical strategy was important for the local union membership and the national union leadership. A national strike, supported by the UAW executive, would have been illegal under US labour law. GM knew this and went to court to convince a judge of the illegality of the action. If the firm had been able to convince the judge then the UAW would have been accountable for GM's losses, running into billions of dollars, and effectively bankrupting the union. However, throughout the strike action the UAW maintained that what was at stake were *local* workplace terms and conditions. In establishing the geographical focus of the strike in this way

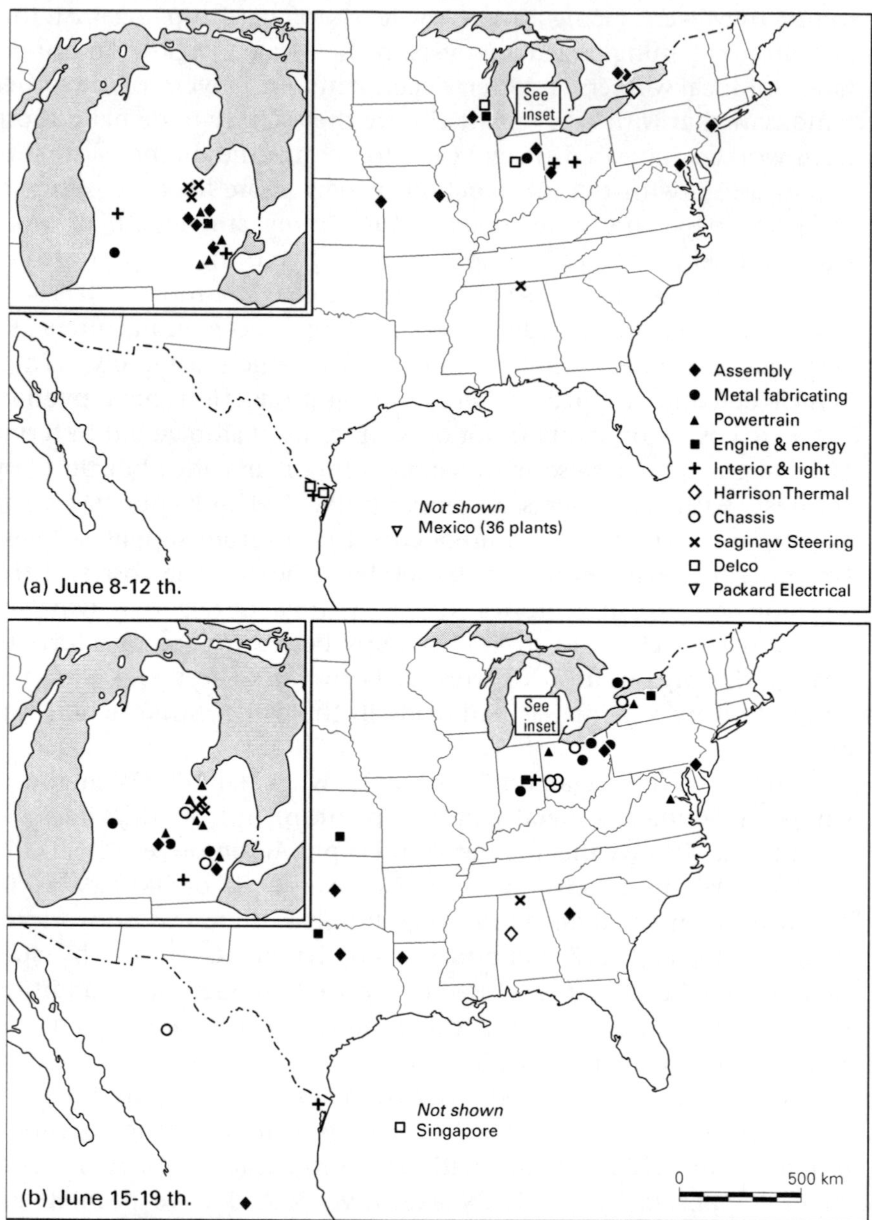

Figure 6.1 Location of GM plants first affected by the VAW strikes
Source: Herod, 2000c: figures 1 and 2

the UAW were able to turn GM's international production system to its advantage. By targeting important production sites in the company's wider corporate network of production and distribution the effect of actions was amplified. So, as one workplace stopped production so there were knock-on effects running right along the network. By the time the strike action ended local workers had extracted a number of concessions from GM:

- The organization of work would not be changed in the two workplaces.
- 180 million dollars would be invested in the metal stamping plant, the first to go on strike, in exchange for a 15 per cent increase in productivity.
- GM would withdraw its complaint that the strikes were illegal, meaning that UAW would not have to go to court and risk fines possibly of billions of dollars.
- Both Flint plants' futures were assured in the short to medium term (source Herod, 2000c).

In sum, this example demonstrates how workers acting alone in place are able to exert their agency to good effect. Arguing that the strikes were over local issues, such as workplace health and safety and overtime payments, the UAW were able to disrupt GM's production systems. The union's Flint workers turned the widespread use of JIT manufacturing by GM across its transnational operation to its advantage. GM relied on the Flint workplaces for components central to the production of its most profitable products, namely pick-up trucks and sport utility vehicles. Choosing strategically important places in GM's global production network, and stressing that local issues were behind the strike allowed the UAW to gain the upper hand, economically and politically, over GM. Specifically, by interrupting GM's finely balanced network of production and component supply, the Flint workers used GM's JIT system against it. Despite appearing to have the odds stacked against them – working in a place already hit hard by industrial restructuring and employed by a large 'mobile' multinational – workers at the two Flint plants thus managed to use their knowledge of the firm's organization of production to achieve real concessions.

WORKERS AND LOCAL COALITIONS

Though the Flint case is one of countless examples of local worker agency, it is equally common for local workforces to act in tandem with other place-based or place-bound actors. These local alliances or coalitions allow workers to reach out from the workplace. They link labour issues with wider questions about the present condition and future character of towns and cities. And they make the agency of wage workers potentially more effective by drawing-in the resources and voices of other local groups. We can think of place coalitions as particular groupings of various place-based and place-bound actors who make a strategic decision to join forces over a given issue or set of issues. In effect, these coalitions pool local agents' various power repertoires, leading (hopefully) to a whole that is greater than the sum of the individual parts. At the most general level, place coalitions usually form because of a real or perceived 'trigger'. 'Negative triggers' are those that seem to pose a threat to a given locality (for example, when a firm proposes to close down a large factory or relocate production elsewhere). 'Positive triggers' are those where local actors seek to exploit an opportunity (for example, to attract new forms of public or private investment that could go to one of several different places nationally or transnationally). Clearly, what can seem a negative trigger for one place can appear to be a positive one for other places. Sometimes there is no particular trigger at all, just a desire – led, perhaps, by a few key local individuals or institutions – to improve the quality of life (economically, socially, culturally, etc.) in a particular place. As mentioned earlier, it is the common experience of living in a locality that impels otherwise quite disparate local actors – be they a local government, a branch-plant of a TNC, a chamber of commerce, or a local religious organization – to come together once triggers have been recognized or where a there's simply a general desire to boost local fortunes.

Clearly, in a world where many firms can choose to (re)locate in a range of places, different local coalitions frequently compete with one another for jobs and investment (refer back to the discussion of competitive production relations in Chapter 4). This competition will not be limited by national boundaries, though it may often involve place alliances in the same country vying with one another. Competition to retain existing, or to attract new, inward investment takes place at a number of geographical scales: metropolitan, national, macro-regional and global. For example, Delhi might be in competition with Bangkok for the re-location of a large call centre from Glasgow; equally, Rio de Janeiro and São Paulo might lobby the Brazilian government for new development grants. Thus, Jonas (1992: 350) has argued that:

> a class politics grounded in community has given way to a community-defined territorial politics of location in which rival place-dependent growth coalitions compete to attract globally mobile capital to their communities.

It is important to note that this place competition does not simply pit 'mobile firms' against 'immobile workers'. Those myriad place dependent groups who are not wage workers – families, local shops, small firms, local governments and so on – also obviously benefit from successful attempts to secure inward investment. This is precisely why place coalitions form in the first place: *diverse* local actors see the *common benefits* of taking an active role in promoting their locality over others elsewhere.

This said, it would be wrong to think that there is simply one coalition that, in any given time and place, can represent the whole local community. Within localities it is also possible for more than one coalition to form. No one coalition can as a matter of right claim to represent the views of all local interests. Coalitions and counter-coalitions might form. Owners of the means of production – the local business elite (powerful local economic and political individuals), for example – tend to be best placed to shape the agendas of place-based coalitions. An example of this occurred in Manchester, England during the 1980s and 1990s. During those decades Manchester was a de-industrializing city rife with unemployment and very much seen as a 'grimy northern city'. Led by Graham Stringer and Bob Scott, an energetic, ambitious and focused coalition emerged that tried to 're-brand' Manchester. The coalition was responsible for the city putting in bids to host the Olympic and Commonwealth Games, and for aggressively seeking out British and European governmental aid for the city region (see Peck and Ward, 2002). However, these 'elite coalitions' do not always form unopposed. Depending on local circumstances, other groups and organizations might have different ideas for how they would like their place to evolve. Whether they can rally enough local support to have their agendas taken forward is a contingent matter. In sum, not all coalitions are made up of the same sets of interests, or seek to make the same sort of difference in the same sort of way.

Specifically, we can identify three types of place-alliances that involve local groups of wage workers (Figure 6.2):

1 Cross-class alliances: here local labour joins forces with employers, local government and other elite actors to defend existing investments in place or attract new ones (Harvey, 1985). In Jonas's (1992: 350) words, 'place competition has led to a paradoxical situation in which locally dependent class actors take up political positions in opposition to what might be expected from their objective class interests'. As the Manchester case illustrates, cross-class alliances often take the form of what are called *growth machines*. These form and

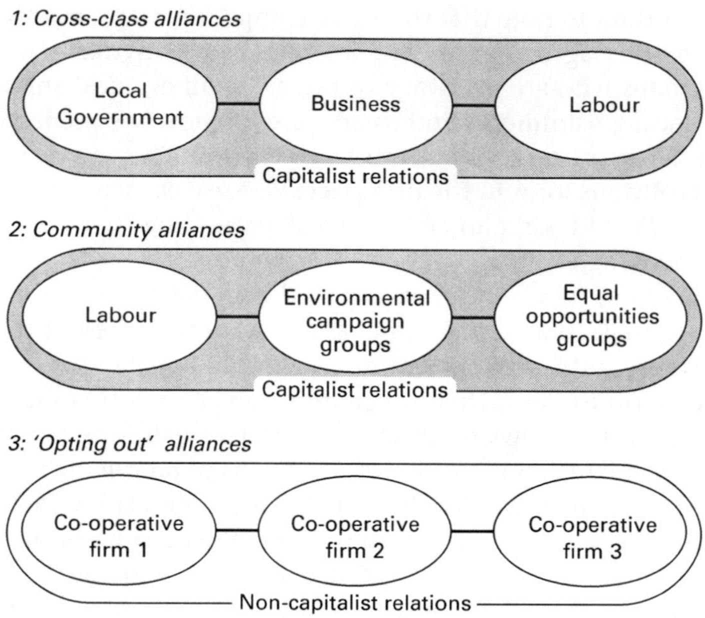

Figure 6.2 Three types of place-based alliances

organize to protect exchange values in place, that is the money that can be made from the renting of houses or factories, or the development and selling of property. Locally embedded firms and regulatory bodies try to intervene 'directly in the local economic development process in order to protect, enhance, or create a context of exchange linkages that will benefit them' (Cox and Mair 1988: 308). As the sociologists who first developed the concept, Logan and Molotch (1987: 32), explain:

> They might quarrel among themselves over exactly how rents [profits] will be distributed among parcels, over how, that is, they will share the spoils of aggregate growth ... [but] virtually all place ['locally-dependent'] entrepreneurs and their growth machine associates ... easily agree on the issue of growth itself.

Of course 'growth' can mean different things to different local interests. It might mean economic growth or it might mean social development in the form of reducing income inequalities. While these are not necessarily mutually exclusive, in the sense that both of these can be pursued at the same time, under global capitalism economic growth – that is, the making of more profits for those that own land and properties, whether residential or business dwellings – tends to be the norm.

Where does labour in place fit within these place coalitions? Logan and Molotch (*ibid.* 81–2), argue that:

> Although they are sometimes in conflict with capitalists on other issues, labor union leaders are enthusiastic partners in growth machines ... Union leadership subscribes to ... development because it will 'bring jobs'.

Organized labour, in the form of local unions, may work with locally dependent business, government and other organizations to promote local economic growth. Of course, not all workers are unionized. Nor do all unions work willingly with local capitalists. As we shall see in Chapter 9, in some cases workers are able to think beyond place. They acknowledge that what might be good for them where they live and work necessarily involves a deterioration in the livelihood of fellow workers elsewhere.

2 Community union alliances: though it may appear that few local actors would oppose growth coalitions, it is important to ask the questions: 'growth for who?' and 'what are the consequences of growth?' Given that most towns and cities are complex amalgams of working and non-working groups, it is always possible that some local actors stand to lose – or to at least be ignored – in any growth alliance. Workers may be one element of an alternative coalition organized in place to promote a different vision of local economic development. Others may join with them (for example, environmental organizations, gay pride groups, anti-racist NGOs or other single-issue bodies), mobilizing around issues where what is at stake impinges directly upon their own interests. As Cox and Mair (1988: 314) explain:

> Troublesome resistance to [growth] coalition strategies arises not only from class-conscious workers. It may also appear in the form of local social movements rooted outside the workplace. Many of the local movements that contest the projects of business coalitions practise displaced forms of class struggle organized around living place and neighbourhood issues.

When, as they frequently do, trade unions represent workers in these types of grass-roots alliances the result is what's termed 'community' or 'reciprocal' unionism'. Wills (2002: 468–9) explains it thus:

> Community unionism is about finding common cause between unions and those groups cemented around affiliations of religion, race, gender,

disability and sexuality, with those providing a particular community service with those fighting for a particular cause.

Here workplace issues are explicitly linked with wider local community issues. Thus, if the alliance aims to attract new investments these will not just be industrial ones but perhaps also cultural or educational ones that have wider community benefits. Here, then, the aspirations of local civil society at large – workers, retirees, dependents, the unemployed, men, women and children, etc. – are emphasized, so that issues inside and outside of the workplace have wider resonance. For example, in the Los Angeles garment industry community organizing has recently involved linking the exploitation experienced by Latino and Asian workers everyday at the workplace with the broader agendas of the immigrant communities (Bonacich, 2001). Though community unionism might, in some circumstances, be regarded as a 'cross-class' phenomenon, it is better understood as (i) building intra-class bridges (specifically, between class fractions and occupations within a local working class) and (ii) building bridges between class and non-class identities and concerns. It is, in other words, a very ecumenical thing.

3 **'Opting out' alliances**: here a number of local civil society groups and local workers act in concert, stepping *outside* the capitalist system altogether. They organize alternative non-capitalist ways of trading goods and services and of utilizing workers. Local capitalist firms, for obvious reasons, rarely join such alliances. Opting out coalitions are relatively rare. Capitalism is today such a 'normal' framework for conducting economic affairs that few people think it possible to live and work under a different system. Consider how you would go about living without selling your labour power in return for money and then using that money to buy goods and services to survive. Opting out is only viable when enough people in the same time and place agree to act together to make a non-capitalist existence possible. The logic underpinning the strategy of dis-embedding or de-linking from global capitalism is that by acting within it workers and those who dependent upon them are, in effect, reproducing it. For the most part those organizing labour or others in place have tended to mobilize inside capitalism in order to change, rather than to reject, the terms under which workers must labour. This is because despite often finding themselves in opposition to business, labourers derive their livelihoods, and often their sense of identity, from the work they do. So there is a paradox: how to improve the conditions under which workers perform their tasks,

how to maintain their sense of identity, but also 'wiggle' so much that they opt out of the global capitalist system? Opting out is not, we should note, the same as dismantling capitalism. That would require a revolution involving different workers in multiple places. So though a relatively rare act of worker agency, opting out of capitalism at the local scale is far easier than abolishing it altogether as a global system. The decision to opt out is made either when a place has suffered serious economic decline relative to the myriad other places it is related to, or when an ethical stance is taken by a local community on capitalism and its ills.

In practical terms, opting-out coalitions typically have three aims. First, they aim to make goods and services in such a way that the whole community benefits, not just a few owners of productive units who 'cream off' the profits. Second, they organize local firms such that workers at *all* points on the occupational ladder have good pay and conditions – partly because the workers might be made the collective owners of those firms. Finally, though opting-out coalitions do not necessarily cut off their place from the wider capitalist world economically speaking, they often aim to get more control over their economic fortunes by becoming self-reliant in the basic commodities needed for daily reproduction. It is important to note the difference between communities who have voluntarily opted out and those who pursue non-capitalist economic activities because they have little choice (Box 6.2).

Box 6.2 Non-capitalist ways of working and living

For a variety of reasons some people are not part of the formal economy. In some cases this is a deliberate decision to opt out of the capitalist system. Ideological or political beliefs in alternative – non-economic – ways of working and living have led a growing number of groups or networks to form and to exchange one good or service for another. An example of this trend is the number of Local Exchange Trading Systems (LETS) that have formed in the UK in the last two decades. LETS are 'examples of local currency systems through which participants can accumulate and expend money for a range of tasks and services' (Leyshon and Lee, 2003: 32). Rejecting the exchange of money for goods and services LETS operate through members trading directly with each other. So one member may do some housework for a second member and in return will receive a 'credit' – that the two members will agree – or in return the second member will perform a task for the first, such as washing the car or collecting the shopping. And so exchanges take place between members of the LET that are not assigned a monetary value, as in the capitalist system, and so there is no profit margin. Rather members barter with one another. Since the formation of the first LET in the mid-1980s they have grown in number and in size. A committee elected by members governs each LET. This oversees exchanges and ensures members remain inside the legal framework. In exchanging goods and services in this way LET members are actively rejecting the capitalist exchange mechanism and actively constructing an alternative economic system. This critique of capitalism is confirmed in recent research into the LET in Stroud, in the South West of England, where it was found that 31 per cent of members had joined for 'ideological reasons' (Aldridge et al., 1999).

In other cases some social groups may not be able to join the formal economy. Their opting-out maybe less voluntary and may include, among others, the short- and long-term unemployed, the unskilled and poorly educated, people with criminal records, and those with mental and behavioural problems. Though efforts may be made to get these people back into paid employment (for example, through government training schemes), sometimes the only alternative to poverty or living on state welfare payments is to create less formal and in some cases non-capitalist economic institutions. An example of this type of activity is Work Force Inc. which is based in the Latrobe Valley, outside Victoria, New Zealand. Formerly a 'hot spot' of energy generation, between 1986 and 1996 over 6,000 jobs were lost in the electricity, gas and water supply sectors. Economic restructuring was accompanied by a more widespread undermining of the local communities. Unable to leave the community, the young and the old were left to make do in the Valley. Work Force Inc's niche is 'between the formal job market and available services to the unemployed' (http//www.arts.monash.edu.au/projects/ cep/index.html). The aim of the organisation is 'to create healthy, cooperative and empowered local communities, where social and economic goals intertwine. Local development is fostered through a people-centred, grass-roots approach where facilitators take on an enabling and responsive role in assisting individuals to bring ideas to reality.' This cooperative widens out the meaning of 'economy' to include non-capitalist goods and services

Clearly, the three categories of place coalition described above are ideal-types. Unsurprisingly, each of these types of place-bound coalitions take on a particular local form, depending on the context in which the coalition finds itself and the types of strategies (geographical and others) it pursues.

LABOUR WORKING IN PLACE WITH OTHERS

Having explained the principal types of synergistic agency that are, in theory, available to wage workers, we turn now to some illustrative examples. First, we consider the case of Clarksburg, West Virginia where local plant workers, local employers and local government joined forces to resist the closure of the local glass manufacturing plant. Then the Baltimore Living Wage (BLW) campaign is discussed to show how reciprocal union coalitions operate. Third, we use the example of the Mondragon cooperative, in the Basque area of Spain, to highlight how it is possible for social groups to 'opt out' of the capitalist system. Drawing upon these case studies, the aim is to emphasize synergistic agency – labour acting in concert with local others – and the possibilities available to workers to resist the real and imagined threats of hegemonic despotism.

In these and other cases of local alliance action, it is important to note that the idea of 'community' becomes a potent resource and something that is struggled over. Despite the fact that the various constituent parties in local alliances are all place dependent in some way, this does not automatically lead to coalition formation. Rather, coalitions form because the constituent parties believe that what is in *their* interest is also in the interest of *all* local coalition members. The idea of community can be vital in defining what the 'local' interest is: it serves as the 'glue' that brings and binds coalitions together. As Cox and Mair (1991: 206) explain:

> Local interests have to be [actively] defined ... Identity is local: it is in terms of local community, its history, its future, that people are asked to understand themselves. It is local things, events, personalities, institutions that people are suppose to take pride in.

Community is a complex term. It has a social and a geographical meaning. Socially, it names 'insiders' and 'outsiders': that is, those who 'belong' and do not belong to the community in question. Geographically, ideas of community often have territorial referents – they refer to named locations. This is why Fitzgerald (1991) considers local alliances to be 'community-defined movements'. The members of the alliance actively construct, and then agree upon, an idea of who and what the community is and what its interests are. In effect, they concoct an idea of what their *place* is or should be all about. This is what the cultural critic Edward Said (1978), writing in another context, called an act of 'imaginative geography'.

Clarksburg, West Virginia: a cross-class alliance in action

First we turn to Clarksburg in the state of West Virginia, a small city with a population of 22,000 that up to the 1960s was the site for five major glass production plants. By 1977 all but one of the plants in the city had been closed, as over one in four state workers in the glass industry lost their jobs. In November 1978 the Brockway Glass Company announced it would close Clarksburg's last glass plant in March the following year. It was estimated that the closure would remove US$12 million from the local economy, cost the city government US$70,000 a year in lost revenues and make 1,300 workers redundant. Local businesses, city and state government and local unions stood to lose in different ways if the Brockway Glass Company went ahead with their decision to close the plant. In Herod's (1991: 393) words, 'maintaining the [plant] became a civic priority'.

In response to the proposed closure a local coalition consisting of the Harrison County Association for Industrial Development (AID), the Clarksburg Chamber of Commerce, the city government and a number of local businesses formed. It lobbied for different pots of money – national, state and local grants – that could be used to attract a buyer for the Clarksburg plant. Although not part of the formal coalition, the plant's American Flint Glass Workers' Union (AFGWU) worked with it in identifying the Anchor Hocking Corporation (AHC) – a Fortune 500 company – as a potential purchaser. The Corporation was struggling to meet the demands for its products and had begun to look at ways at expanding its production capability quickly and cheaply. It appeared that all parties would gain if the Anchor Hocking Corporation would purchase the plant.

On the one hand, it was clear that for Clarksburg's future economic sustainability it was important that the plant be kept open. The city had already suffered a series of plant closures. Unemployment was rising and local economic fortunes were on the wane. The Anchor Hocking Corporation was able to use this locality dependence of the coalition as a point of leverage. It argued that the plant's unprofitably meant the cost of modernization would only be financially viable if the corporation was loaned money by the city and state governments at low rates of interest and if the unions would agree to lower wage costs and changes to the organization of work. On the other hand, the corporation was itself under pressure to make adjustments to how and where its production occurred. And yet the corporation was able successfully to downplay the gains that it would accrue through buying the Clarksburg plant. Instead the context for negotiations was set by the relative immobility or locality dependence of the coalition. While the corporation was able to present itself as highly mobile, able at will to withdraw from negotiations, the coalition of place-based and place-dependent interests was unable to distance itself from its own immobility. As a result, although the cross-class alliance was successful, in that it convinced the Anchor Hocking Corporation to take over the plant, this success was at a cost: US$8.5 million of public grants and a decline in the terms and conditions of workers.

In sum, this example has shown how labour acting alongside others is able to defend its place-based interests (see also Box 6.5). Workers at the Clarksburg glass plant were able to defend their jobs through forming an alliance with those outside of their own class. Local employers and workers came together with local government, overcoming temporarily their class differences and protecting their own interests.

Baltimore's Living Wage Campaign: an example of reciprocal unionism

For our second case study of local coalitions we move to the east coast of the United States and to Baltimore. Another US city whose economy has suffered during the last two decades, Baltimore was one of the first places to introduce the waterfront model of urban redevelopment (Harvey, 1989b), which has subsequently become widespread in both the United States and Europe. This focused on renovating the Baltimore inner harbour through the construction of expensive apartments, luxury hotels, sports stadia, and retail centres. Pursuing this particular local growth model meant that relatively well-paid and secure jobs in manufacturing were replaced with low-paid and precarious service sector employment, (as, recalling one of the previous chapter's case studies, has happened in the Southeast of England). This model was supported (and promoted) by a local coalition of businesses and government – led by Mayor Schaeffer. It argued that Baltimore had little option but to reinvent itself as a place where tourists could stay and play. So the coalition set about competing aggressively with other places for inward investment, poaching firms from elsewhere with generous subsidies and tax incentives.

However, another coalition was formed to campaign against some of the consequences for workers of the pursuit of this model of redevelopment. Baltimore's living wage campaign is an example of what can be achieved when labour and community organizations unite around a common cause. An alliance formed between one of the largest public sector unions in the state of Maryland, the American Federation State County and Municipal Employees (AFSCME) and Baltimoreans United in Leadership Development (BUILD), an association of 50 mostly African-American churches. Both groups were concerned about the growth in low-paid jobs in central Baltimore. For the union, the privatization of public services meant they were losing members: for the churches, their congregations were declining. Relatively well-paid African-American families were moving out of Baltimore city and into the suburbs. BUILD organized public demonstrations to resist what it saw as the declining work and living conditions experienced on a daily basis by its communities. It soon realized that in order to change the lives of its members who were performing low-end 'shit jobs' it needed to get inside the workplace. As the then lead organizer of BUILD, Arnie Graf, argued 'a lot of our organizing became pushing government to organize subsidies to supplement the downward spiral of people's wages ... We moved into the arena of low-wage work workers in the city because that's our constituency. That's who lives around the churches and the neighborhoods where we are' (quoted in Walsh, 2000: 1601).

AFSCME and BUILD organizers targeted the low-wage workers who undertook contract-only jobs for the city government, arguing for a 'living wage' that would ensure that no family of four with one full-time working adult would live below the poverty line. Despite opposition from local firms the measure was passed in December 1994. The pay of service workers contracted to city government was to rise to US$6.10 in July 1995 (when the US federal minimum was US$4.25 per cent), rising to US$7.70 over four years. Over 1,500 jobs, such as school bus drivers, janitors and fast-food service workers were covered by the deal. While not covering all those who labour daily for a wage that keeps them at the bottom of the labour-market, the introduction of a living wage in Baltimore did improve the pay for some of the most poorly paid workers in the city.

In sum, this example shows how workplace issues can be made to matter to a wide audience. The introduction of a living wage in Baltimore improved the working lives of the city's service workers. These workers also had lives outside of the place of work. Through the connecting of the lives of workers inside and outside of the workplace a basis was provided for the formation of a reciprocal coalition (Walsh, 2000). It was able to draw upon each group's power repertoires and to extract concessions from the pro-growth coalition that already existed in the city. Such living wage campaigns in the United States have served as inspiration for the low paid in other places worldwide.

Mondragón: working outside of capitalism

Our third and final case study of place coalitions is the Mondragon Cooperative Corporation (MCC) that was formed in the Basque area of Spain in 1955 with the aim to 'provide Basque workers with good stable employment and to contribute to the growth of the Basque economy' (Clamp 2000: 559). Unhappy with the close relations that existed between large businesses and the Spanish government in the aftermath of the Spanish Civil War, Father José María set up the first cooperative as a means of establishing a degree of economic and political independence for the area. Since its creation the cooperative has spawned other worker cooperatives, a bank and a number of educational establishments. In this case, ethnic relations grounded in a particular place at a particular time took precedence over potential class differences. Individuals were held together in the cooperative by their own sense of Basqueness, and their opposition to the Spanish national government. Class differences, such as those between professional bankers and lower-paid workers, were subsumed to a politics of shared ethnic identities.

Since its formation the MCC has expanded what it does. It is now a federation of 85 worker cooperatives and 22 firms that are wholly or jointly held by the cooperatives, employing in total over 34,000 workers, of which over 65 per cent are Basque residents. Cooperatives are associations or groups of associations that supply goods or services and whose profits are fed back to members. The MCC's founding principles were:

- Worker sovereignty
- Worker participation in the management of cooperatives
- Cooperation between cooperatives

Work was to be organized along the following non-capitalist, more 'just' lines:

- Profits were pooled between the different cooperatives.
- Surplus workers were offered employment at other cooperatives.
- Flexible calendars were introduced to accommodate flexible workloads.
- Loans were given to individuals and cooperatives on discounted rates.

From its formation until the 1980s the federation of cooperatives were able to protect themselves from the economic pressures experienced by their non-cooperative competitors. Through fostering inter-cooperative working, and supporting new cooperative start-ups the federation was able to achieve growth while sticking to its principles of fair and just employment.

Since the early 1990s, however, the MCC has found its principles under increasing pressure. Despite operating 'outside' of the capitalist system, it has nonetheless been affected by the changes in the global production system (see Chapter 4). As the geographical mobility of capital has increased over the last few decades, so competing against those inside the capitalist system whilst remaining outside of it has become increasingly harder. The MCC faced a direct challenge to its very organizing principles: would it be able to stay true to its philosophy of cooperation while still being able to compete against other businesses, some of whom were prepared to pay lower wages and offer workers a less comfortable working environment?

The answer is a mix of 'yes' and 'no'. The MCC looked at how it organized production and decided to make a number of changes. Traditional methods, which tended to be relatively labour intensive, were gradually replaced with newer methods that required fewer workers and made more use of new technological developments. In addition, the MCC shifted the emphasis in its growth strategy away from new start-ups – which has been the basis of its early expansion – and towards the formation of joint ventures, often with non-cooperatives. Hence, despite continuing to operate outside of the capitalist system, the organization of the MCC and its activities were not unaffected by some of the tendencies inside it. Due to forming joint ventures the federation became an employer outside of the Basque country, forcing it to rethink its central objectives. Its principles became (i) to preserve MCC jobs, (ii) to maintain the profits produced by MCC firms, (iii) to comply with local labour practices in overseas operations and (iv) when possible to shift production from overseas back to the Basque country (Clamp, 2000).

In the last 10 years the experiences of the MCC 'experiment' demonstrate how hard it is to operate outside the clutches of capitalism. Workers in Thailand and in China labour under worse conditions and receive lower pay than those in the Basque country, and yet the MCC has been forced to use these workers to remain competitive. They, of course, are unable to share in the ownership and the profits with their Basque 'colleagues'. In sticking to their 'local' principles, MCC has created a sizeable insecure workforce. Even it its homeland it has not been completely successful in protecting the conditions under which its workers labour. Lower-skilled jobs, such as security guards have been contracted out to private firms, as the cooperative has striven to balance its egalitarian principles with being able to generate profits for its members.

Despite their just beginnings cooperatives, such as MCC, which strive to organize 'outside' of global capitalism have found themselves under increased pressure to contravene their own principles. The options are stark ones: keep production in the Basque country, where pay rates are higher, and risk job losses and firm closures, or establish joint ventures in those regions where pay rates are low. As Clamp (2000: 575) explains, 'the cooperative business cannot readily address the problems of labour solidarity ... as long as cooperative firms must rely on global markets'.

FROM 'PROGRESSIVE' TO 'REGRESSIVE' LOCALISM (AND BACK AGAIN)

In different ways, the four case studies discussed in this chapter demonstrate amply the agency of labourers at the local level. Yet, those on the Left have not traditionally been convinced of the concessions labour can extract from business and regulatory authorities through exercising its agency in place. For example, Harvey (1993: 24) has argued that 'place-bound politics appeals, even though such a politics is doomed to failure'. While this might be the case in some localities, it is certainly not the case in all. Indeed, it should be clear from the examples in this chapter that under certain conditions workers either acting alone or alongside others can put employers and regulators on the back foot. In one sense, and as this chapter has argued, it is possible to point to 'progressive' localisms. Workers and other members of local communities can really improve their lot. The 'difference' made might involve a change in conditions inside or outside the workplace: a safer place of work and better hourly rates or better quality public spaces and lower levels of pollution. As we saw in each of the four examples, 'pure' or 'synergistic' forms of worker agency were able to extract gains from business or government.

In the case of the workers in Flint, Michigan, they secured new investment in their plant, and a more secure medium-term future. In Clarksburg the workers joined with other locality-dependent interests to stop the glass plant being closed. In Baltimore the pursuit of a particular model of redevelopment led to the creation of a low-income working poor 'community'. Finally, the Mondragón Cooperative Corporation (MCC) emerged out of a commitment to the well being of the local community and a rejection of the way capitalism generates and distributes profits.

However, it is also possible to offer an alternative explanation of what appears to be evidence of 'progressive' localism. By definition, organizing in place means that one geographical scale is privileged over all others. What is good for place-dependent groups is defined at the local level in abstraction from other geographical scales. And yet, as we explained in Chapters 3 and 4, we live in an inter-dependent world. Places are not bounded locations, but are part of relational networks and flows. The agency exerted by labour and other groups in one place may change the parameters within which labour can 'wiggle' in other places. In exercising its agency locally, labour might actually make things worse for others in the same economic class in other places. Whether or not a particular strategy is 'progressive' or not depends on where, and at what geographical scale one looks. In the case of our first example, Flint workers made gains through the agency of the local union, the UAW. The outcome of the strikes might have had more 'regressive' outcomes in GM's other plants, where workers were temporarily unable to work because of the Flint strikes. Indeed, non-Flint workers may have been laid off, as the Flint plants stopped their production of necessary parts. And the same goes for Clarksburg, Baltimore and Mondragón. In the case of the living wage of Baltimore, the higher rates paid to city government workers might have meant that less use was made of them. In their place more use might have been made of non-city government workers, with these workers hired on worse terms of employment. In effect one group of labourers subsidized another.

Why is this important to us in trying to understand the progressiveness or otherwise of the agency of labour in place? Well, it should be clear that what, at first sight, appear to be examples of the 'progressive' possibilities open to labour through exercising its agency locally need to be interpreted in the context of other changes that may occur as a result of wage workers' actions. It may be, for example, that 'progressive' localisms and 'regressive localisms' coexist, that the effects of in-place labour agency on geographically distant but inter-connected *spaces of work* may be a complex blend of gains and losses. More fundamentally, however, there might be something inherently 'regressive' in organizing in place, that any gains made by labour are by definition off-set by the ability of businesses to displace the costs on to other workers elsewhere. We return to the issues of progressive and regressive strategies in Chapter 9 in a discussion of competing definitions of justice.

SUMMARY

This chapter has set out how workers, organizing inside and outside of the global capitalist system, have improved their lot, often in coalitions with others. This is not to claim that these small concessions are not won without a cost. Sometimes the cost is in the form of a trade-off between holding onto employment in a place and the conditions under which labouring occurs. In other instances the cost falls elsewhere, in the same country perhaps, or in another part of the world entirely. From the examples used in this chapter then it should be clear that what is progressive in one place may be regressive in another. Fighting for and winning better terms and conditions of work in one place might lead to the degradation of employment surroundings in another. Quite simply, one person's progressive localism might be another person's regressive localism. In Part Three we turn our attention away from workers' local actions to consider translocal forms of agency. Increasingly, moving across space (migration) or organizing over space (creating cooperative inter-place connections) are important geographical strategies for wage workers in the twenty-first century.

Part Three

RE-SCALING LABOUR

seven

Dis/placing Labour

In Part Two we explained and illustrated the difference that place makes to workers in the contemporary world. The place-based, and sometimes place-bound, nature of workers' existence can, as we have shown, put them on the back foot (Chapter 5) or else compel them to take a proactive stance in defence of their home places (Chapter 6). In Part Three we wish to change the geographical focus. As explained in Part One, a geographical strategy that workers can employ to improve their life conditions is to 'up-scale' their actions and command space. In this and the following chapter we explore the two principal forms such a strategy can take: namely, labour migration (movement *across* space) and inter-place worker cooperation (forging links *over* space). In both cases, workers can and do up-scale their actions at a variety of levels, from the national to the global, and for a variety of time periods. This up-scaling emphasizes the point we made in Chapter 4 that geographical scales are socially constructed rather than given in nature. Once they take on a certain material solidity, these scales can both enable and constrain the actions of different groups and institutions. In Part Three, following on from the last chapter, we again emphasize worker agency – that is, workers' capacity to exert some control over their destinities. However, as we shall, labour migration is not always about the voluntary exercise of worker agency. Rather, it is an ambivalent phenomenon, wherein many workers leave their original home places (temporarily or permanently) because they have few other options than to do so.

At first sight, it may seem odd to devote a chapter to migration in a book on work and wage-workers in the modern world. This is because the nature and causes of migration extend far beyond those surrounding labour migration alone. Consequently, many contemporary labour researchers do not discuss migration at all (see, for example, Herod, 2001). However, while labour migration is just one element of migration and often hard to disentangle from other types of migration, it is nonetheless absolutely central to the concerns of this book. First, it is a major means whereby many workers can 'escape', for a greater or shorter period of time, the place-based nature of their normal, everyday existence, with all that this entails. Second, migration brings workers in one place ('source' areas) into direct, face-to-face contact with workers in other places ('destination' areas). Depending on the socio-economic circumstances of the destination areas, this can lead to cooperation or conflict (or mere indifference) among the groups of workers and migrants in question. Finally, recalling the discussion at the end of Chapter 6 on 'progressive' and 'regressive' worker localism, this second point indicates how profoundly *perspectival* the significance of labour migration is. That is, how it is judged on the ground depends wholly on which *specific groups* of workers and migrants are involved in which *particular places* and what *their* needs and wants happen to be.

With these three points in mind, this chapter describes, explains and illustrates the principal types of labour migration in the contemporary world, while exploring the perspectival nature of migration's significance. It is structured as follows. In the first section different types of labour migration are distinguished and a brief discussion of the volume and geography of these types is presented. Following this, we explain how and why labour migration is regulated in our interdependent world. Without regulation, present day labour migration patterns would be very different indeed. Understanding migration regulation is thus the key to grasping which kinds of workers can migrate where and when in the modern era. We then explore the double-sided nature of labour migration in two stages. We look at both forced labour migration, where the 'agency' of workers is reactive, and at voluntary labour migration where workers use physical movement as a proactive geographical strategy. The last part of the chapter then explores the impacts of labour migration on different players in the migration drama: namely, migrants themselves, workers and communities in destination areas and families and relatives in source areas – all in the context of actions by firms and regulatory institutions. A number of case studies are used to illustrate our arguments.

WORKERS ON THE MOVE: THE WORLDS OF LABOUR MIGRATION

Types of labour migration

As we noted above, labour migration is just one type of migration, but labour migration itself assumes many forms, and below we distinguish between several types. These distinctions are far easier to make in theory than in practice, but are nonetheless useful in getting an analytical handle on labour migration:

- **Temporary** and **permanent:** temporary labour migration refers to people that cross international and sub-national boundaries with the express intention of returning to the source area. The Organization for Economic Cooperation and Development (OECD) has an even more precise definition – referring to 'short-term migrants' as those who stay for less than three months. In any case, for such temporary movements of labour, the term *migrant* rather than immigrant tends to be used. In contrast, many people cross international boundaries with no intention and/or no possibility of returning to the source area for any length of time, and therefore can be considered *permanent immigrants*.
- **Internal** and **external:** *internal* labour migration refers to the migration of labour within a given country, while *external* labour migration is the movement of people across international boundaries. Officially, international labour migration means migrating under the rubric of a formal job contract. However, most people can be considered 'labour migrants' or 'economic migrants' (unless they are independently wealthy) so even people who migrate to seek asylum, to study, or to join their family, may then subsequently search for work regardless of their original reason for emigrating. Thus, if we want to discuss the condition of 'migrant' or 'immigrant' workers, perhaps we should extend our definition to all those who are foreign-born, are not nationals of the country of destination, and are employed on a work permit (or visa) basis.
- **Legal** and **undocumented:** we can distinguish between those migrants who have the legal right to residence and/or work in the destination area or country, and the *undocumented* (also called 'illegal', 'irregular', or 'clandestine' im/migrants). In other words, undocumented migrants do not have the requisite visas or permits

either to reside in the country of destination, and/or to be legally or officially employed. Undocumented immigration is defined (in relation to what is legal) by national states. It refers to migrants who enter clandestinely or who over stay their visas in a given national territory, or (less commonly) legal migrants who work informally and are liable for deportation. That is, while many undocumented immigrants enter legally and overstay their visas, many others will be 'smuggled' or 'trafficked'.

- **Skilled** and **unskilled:** there is no clear definition of 'skilled' and 'unskilled'. Migrants bring with them all sorts of skills, some of which are recognized or valued by employers in the destination area and others that are not. However, we can make a broad distinction between low-paid migrants (that is the destination area does not value highly the skills of an immigrant) and professional migrants, such as bankers, doctors, and so on whose work is highly valued by the destination area.
- **Voluntary** and **forced:** voluntary migration refers to people who move because they are seeking higher wages, better employment opportunities, the possibility of learning entrepreneurial skills and/or raising financial resources to begin a business in the destination or source area. *Forced migration* refers to people who are escaping poverty, persecution, and so forth, and have no choice but to move from the source area.

The volume and geography of labour migration

Let us now move on to briefly consider the volume and geography of contemporary international labour migration. Over the last decade or so, what is particularly striking is not so much the volume of migration but rather the diversity of source and destination areas (see Figure 7.1). In addition to movements between former colonies and their colonial powers, people are increasingly moving between countries that have historically had few political, social, or economic links (King, 1995). We can summarize the main movements as follows:

- Migration to the United States, including movements from East, South, and Southeast Asia, Central America, South America and the Caribbean
- Chinese and Indian migrants to Canada
- North Africans, Turkish, Eastern Europeans and South Americans to the European Union
- The movement of Americans, Chinese, Filipinos and Brazilians to Japan
- British, New Zealand and Southeast Asian migrants to Australia

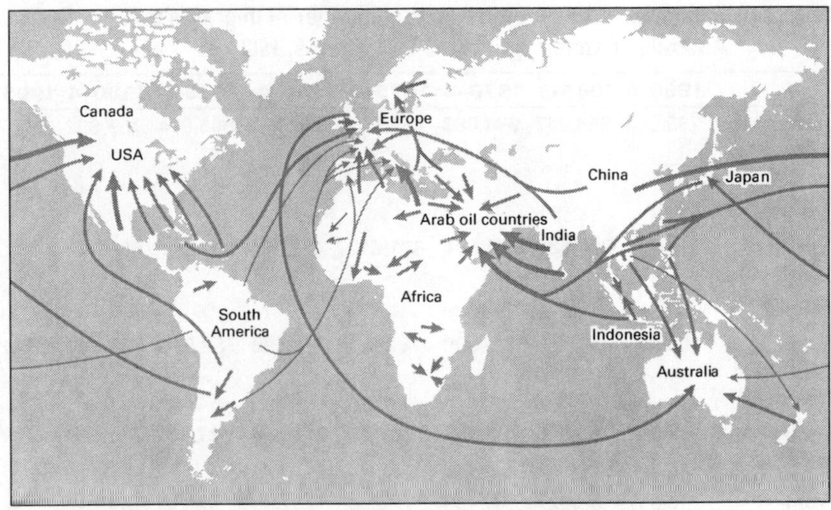

Figure 7.1 Major migration movements since 1973

Source: Based on Castles and Miller, 1998: Map 1.1

- From South Asian and Southeast Asian countries to Singapore and the oil-exporting countries of the Middle East.

We could continue *ad infinitum* about migrant workers criss-crossing the globe. The point of the above discussion is simply to illustrate the diversification of source and destination areas, rather than to exhaustively document the major migrations of workers (for a global review of these movements, see Castles and Miller, 1998, and Held et al., 1999).

Given the above discussion, it would be tempting to conclude that there has been a massive expansion of labour migration across the globe during the twentieth century. In fact, the beginning of the twenty-first century is far from a distinctive 'age of migration' as many observers claim (Zlotnick, 1998). The statistical evidence on migration is often incomparable between countries and more reliable for the advanced economies than for developing countries. However, the available data suggests that the volume of inflows to the former is now only slightly greater than in the 1920s, when migration supposedly reached its pre-twenty-first-century peak. Nonetheless, estimates suggest that some 100 million people are now living outside their country of origin (Hirst and Thompson, 1999). If we look at Table 7.1, we can note that for three major immigration countries, net migration rates are

TABLE 7.1 Estimates of the average annual number of migrants to Australia, Canada, and the United States, 1960–96

	1960–4	1965–9	1970–4	1975–9	1980–4	1985–9	1990–4	1995–9
Immigrants to the United States	283,803	358,947	384,683	459,541	565,007	605,674	769,832	813,730
From developing countries	118,851	200,539	271,684	371,612	481,115	523,047	608,142	633,518
From developed countries[1]	164,591	158,409	112,999	87,829	83,892	82,626	161,690	160,212
Immigrants to Canada	88,008	181,976	158,857	130,127	114,056	137,910	235,509	–
From developing countries	10,857	38,027	67,439	72,299	70,868	97,683	184,547	–
From developed countries	77,151	143,950	91,418	57,828	43,187	40,227	50,962	–
Immigrants to Australia	103,132	134,214	94,521	54,473	79,385	101,550	64,044	102,030
From developing countries	7,298	17,008	26,113	29,239	37,725	63,585	52,102	74,190
From developed countries	95,834	117,207	68,408	25,234	41,660	37,964	11,942	27,840

[1] Developed' countries include the former 'Soviet bloc' countries of Eastern Europe.

Source: adapted from Zlotnick (1998)

positive (that is more people are immigrating than emigrating to the advanced economies) after the 1980s. We can also compare the number and percentage of (legal) workers that are foreign in the labour force of various countries (see Table 7.2). At the same time, official data sources generally miss one of the most significant sources of immigration: undocumented immigration. Indeed, by the very nature of the phenomenon, it is extremely difficult to estimate this type of migration (Delauney and Tapinos, 1998). In any case, whatever data problems do exist, it is clear that significant numbers of workers are crossing international borders seeking work, and this has implications for the livelihood of *all* workers.

TABLE 7.2 The number of foreign workers (thousands) and the percentage of foreign workers in the labour force of selected countries in 1990 and 1998

	1990	1998
France	1,549.5	1,586.7
% of total labour force	6.2	6.1
Germany	2,025.1	2,521.9
% of total labour force	7.1	9.1
Italy	285.3	332.2
% of total labour force	1.3	1.7
Japan	85.5	119.0
% of total labour force	0.1	0.2
Netherlands	192	208
% of total labour force	3.1	2.9
Spain	85.4	190.6
% of total labour force	0.6	1.2
Sweden	246	219
% of total labour force	5.4	5.1
Switzerland	669.8	691.1
% of total labour force	18.9	17.3
United Kingdom	882	1,039
% of total labour force	3.3	3.9

Source: adapted from SOPEMI (2000: 309)

REGULATING MIGRATION

The regulation of migration is achieved at a variety of scales – the macro-regional, national, and local – and by both state and non-state actors. Across the world, social actors at these different scales both *discourage* and *encourage* internal and international migration. An understanding of regulation is essential because regulation serves to constrain or facilitate the geographical strategies of workers. In the following sections, we begin with a discussion of national-level regulation, and then look at the shape of regulation at the macro-regional and local scales.

The role of the national state

Despite all the hyperbole about living in a 'borderless world', national governments actively regulate the *international* migration of workers. Though international boundaries are certainly porous (and national states often have difficulty regulating migration movements), workers are not simply free to move to economic growth areas. That is, the international division of labour is not only the result of *economic* processes, but also *political* processes. One way of conceptualizing this is to use the term 'international labour segmentation' (Samers, 1998), which refers to the way in which national states actively 'sort' workers based on their skills, ethnic and religious background, national origin, and other characteristics. This segmentation will, in turn, have an effect on the spatial division of economic activity through 'brain drains', 'brain gains', and general skill acquisition over time. Thus, to return to the conceptual framework outlined in Chapter 2, we can think of the regulation of the movement of workers in terms of regulatory institutions, regulatory mechanisms, and regulatory objects. In this case, the 'objects' to be regulated are people on the move; the institutions include national governments, but also non-state organizations such as airlines, ferry services, and other private actors; and the mechanisms of regulation include border and police controls, immigration and citizenship rules, and so forth. These mechanisms can then be divided into those that involve the control of entry or exit, and those that concern settlement and citizenship.

But why would states seek to control the movement of workers? After all, would not more liberal immigration policies lead to lower wages (because of an increase in the supply of workers) and thus benefit firms and as a consequence capitalist nation states? Although there are complex reasons why national states encourage or discourage immigration, one of the simplest ways to understand national immigration policies is to view them as the result of a contradiction between the state's desire to ensure economic growth on one hand, and political legitimation on the other. This contradiction becomes acute during periods of combined 'rationality', 'fiscal' and 'legitimation crises' (Habermas, 1976) (see Chapter 2). In this sense, the national state tries to manage acute economic decline, a resulting decrease in its fiscal resources and a loss of public confidence. In relation to im/migration, such crises often involve national governments attempting to appease a Right-wing anti-immigration public, and a pro-immigration public composed of strange bedfellows (such as employers seeking cheap labour, civil rights activists and already settled immigrants). The actions of the French state during the 1980s offer an illustrative example. Key industries in France (such as the automobile industry) sought to resolve their productivity and labour quality problems by shedding 'low-skilled' (primarily Algerian and Moroccan) immigrant workers. Faced with generally rising unemployment, increased welfare payments, the demand by immigrants for longer residence permits, and discontent among the white French public, the French state responded by providing financial and other incentives for immigrants to leave France. In short, the French state experienced problems of accumulation and legitimation, and encouraged return migration as a geographical strategy to overcome these difficulties (Samers, 1999). What can be gathered from the above example is that states are not simply at the mercy of a set of processes commonly called 'globalization' which hinder their ability to regulate immigration. Rather, conflicting national policy goals, nationally-specific human rights considerations, and changing economic circumstances all lead to contradictory national immigration policies.

One might conclude from the above discussion that immigration control is simply about 'First World' states excluding 'Third World' immigrants who are depicted as unwelcome 'others'. However, this is not the only form of exclusion. For example, in what Klotz (2000) calls a 'new non-racial xenophobia' in South Africa, the post-apartheid government is putting considerable effort into restricting immigration from neighbouring countries and clamping down on undocumented immigration. In the same vein, during the 1980s Nigeria expelled mainly undocumented Ghanaian workers as the national economy faltered in the wake of a drop in world oil prices (Van Hear, 1998).

Yet as we noted above, states do not just *limit* labour migration, they actively *encourage* it, and this is true for both countries of immigration and emigration. While the advanced economies have for the most part stopped state-sanctioned low-skilled labour recruitment, they have (since at least the second half of the 1990s) sought to recruit either highly-trained/highly educated workers and/or those with considerable financial resources or entrepreneurial promise. The Canadian government, for example, has developed a sophisticated 'point system' in which one's education, skills, financial resources, or the ability to create employment and contribute to the cultural and social development of Canada can add up to sufficient points for entry (Hiebert, 2002).

In contrast to encouraging immigration, many countries, such as the Philippines and Indonesia, have actively promoted the *export of their own* workers since the 1970s. For example, the Indonesian Ministry of Labour created the *agency* AKAN to assist Indonesian workers wishing to work abroad, and AKAN has worked with the Indonesian Manpower Supplier Association, which is affiliated to many private labour suppliers (*Jakarta Post*, 10 March 2002). Yet why would states encourage emigration when skills at home are valuable?

There are at least two explanations: namely, the hope that many emigrants will return with sought after skills, and (perhaps more importantly) because financial remittances – in other words money sent back home – can amount to very substantial sums. For example, Egyptians abroad remitted US$4.7 billion in 1995 and the Philippines received US$7 billion from remittances in 1996. And between 1970 and 1995, Chinese workers living abroad in Southeast Asian countries had literally *donated* US$3.6 billion to the construction of schools, hospitals, cinemas, libraries, and even highways and bridges in China. Overall, global remittances increased from US$2 billion in 1970 to nearly US$70 billion in 1995 (Bolt, 2000; Stalker, 2000).

Though we have thus far concentrated on international migration, the state also controls *internal* migration. Let us begin by recalling our discussion of different types of states in Chapter 2. These can be used as a general template for understanding how states regulate internal migration. These state types include minimal and interventionist states and they are both compatible with either democratic or dictatorial modes of governance. The advanced economies of the northern hemisphere (and many states in the southern hemisphere) intervene very little in regulating internal labour migration, while interventionist states often use complex policies to regulate migration (typically based on one's place of origin). A prime example is the tight control of rural to urban migration in China (see Box 7.1). Similarly, on the other side of the Pacific, the Ecuadorian government is seeking to limit tourism and thus the migration of workers to the Galapagos Islands because of the effects of tourism on the fragile ecosystem of those islands. In fact, governments regulate internal movements for a complex set of reasons, including the belief that migration to a particular region is the source of higher unemployment, congestion, pollution, lack of housing, and so forth.

Box 7.1 Rural to urban migration in China

Migration to China's coastal cities is not new, but it has grown rapidly since the 1980s. The numbers who now migrate are vast (estimated to be somewhere in the region of 60–100 million, of which approximately 50–60 million migrate to cities). This is not surprising since approximately 20–30% of all rural labourers are unemployed. Traditionally, rural to urban migration was regulated by the *hukou* (or household registration system). This privileged urban dwellers and constructed a complex system of regulations for those wishing to migrate to Chinese cities. The migrants or so-called non-locals (*walai renkou*) find jobs largely through friends and relatives, or follow other rural workers from their own village. That is, each migrant 'group' covers a niche of the urban economy. Circular migration (in other words, the back and forth migration from village to city to village) is common, and often workers return to their native villages during the harvest or spring festivals. Most migrant workers find strenuous and unpleasant jobs in often state-owned and health-damaging heavy industries located on the edge of China's large coastal cities, as well as in informal employment and/or temporary or unskilled jobs in the state and non-state formal sector, especially in the Special Economic Zones and newly-industrializing cities of Southeast China. In particular, young female migrants work in highly repetitive jobs in clothing, consumer goods, electronics, food processing, and pharmaceuticals. Although the *walai renkou* are resented sometimes by 'local' workers, the presence of low-paid migrant workers boosts the urban construction and service industries. Furthermore legal workers and local peasants (especially house-owners) benefit because the former are dependent upon income from renting their rooms, land, and other premises at exorbitant prices in the urban villages where migrants live (Fan, 2002; Logan, 2002).

Macro-regional and local regulation

While national states may be the ultimate arbiter of immigration policy, this does not mean that other levels of government do not have rule-making authority or the capacity for rule enforcement. In the following discussion, we will look briefly at two different levels of governance: the macro-regional and the local or sub-national (see Figure 0.1).

Is there is a macro-regionalization of immigration policy? In other words are macro-regional institutions (state and quasi-state) now dictating immigration policy? Although a battery of international conventions agreed by the International Labour Organization (ILO) and the United Nations (UN) nominally protects migrant workers, not all states have signed these international conventions and their signature does not guarantee compliance. Furthermore, although migration has figured in small ways in the formation of such international trading systems as NAFTA (the North American Free Trade Agreement), MERCOSUR in Latin America (the Southern Cone Common Market) and the ASEAN bloc (Association of Southeast Asian Nations), they possess little legal power in controlling immigration (Pellerin, 1999). Even the European Union (EU), which has moved the furthest towards formulating a macro-regional migration framework, has actually very little power over national immigration policies (see Box 7.2).

Box 7.2 Immigration policy in the European Union (EU)

One way of evaluating the macro-regionalization of immigration policy is to examine the regulation of entry into the EU. Since the 1970s, there has been considerable policy discussion and cooperation relating to immigration between member states of the EU. But it was not until the signing of the Schengen Agreement in 1990 (and its implementation in March 1995), that the EU began to resemble anything like a macro-regional regime with regard to immigration policy. Furthermore, the Treaty of Amsterdam (which came into effect in May 1999) called for the 'communatarization' of immigration matters by 2004. That is, decisions about immigration policy should be subject to the 'community method of decision-making', rather than they being solely the preserve of national governments. But this communatarization is not quite macro-regionalization. Why? The EU institutions responsible for decision-making are the European Commission and the Council of the European Union (commonly known as the Council of Ministers). The Council of Ministers dominates the commission in terms of decision-making, with 'unanimity' the mode of passing legislation at least until May 2004. In contrast to the role of the commission and the Council of Ministers, the European Parliament (EP) is limited to a consultative role during this period. Similarly, the European Court of Justice (ECJ) has been granted only limited jurisdiction over ruling on the way in which EU institutions interpret European Acts. The ECJ has no jurisdiction with regard to national measures adopted in relation to the crossing of borders (internal movements) largely because of security concerns on the part of national governments. Furthermore, the Commission and the EP are effectively marginalized, while the requirement of unanimity in the Council of Ministers acts as a formidable obstacle to decision-making. In short, although decisions on immigration policy are scheduled to be brought completely into the European Union by 2004, they remain subject to the community method which involves inter-governmental (that is international) disputes. In other words, what appears to be macro-regionalization is really communatarization (for a complete discussion, see Geddes, 2000). The EU, despite appearances, thus currently lacks a macro-regional immigrant policy that it effectively applied.

Our interest in *local* (or other sub-national) regulatory bodies (such as municipalities, provinces, regions, or American states) is not so much because they directly control entry and exit (these decisions are normally left to national states), but because they regulate settlement and citizenship. In other words, they can strongly influence the volume and nature of labour migration into their localities. Below we look at one example of sub-national regulation (that is at the level of a US state).

California's Proposition 187 In the context of recession, but especially job cuts in the hard-hit post-Cold War defence industry and shrinking state tax revenues in early 1990s California, policy makers in the state legislature argued that something had to be done about undocumented immigration. Indeed, they insisted that undocumented immigrants cost the state of California US$5 billion of tax-payers' money, and that the state had become a 'welfare magnet' – that is, it attracted poor migrants because of its relatively generous welfare benefits. Policy makers further maintained that in order to prevent the spread of disease, alleviate overcrowding in schools, and reverse declining wages and competition for scarce jobs between low-skilled native Californian workers and immigrants, some sort of legislation had to be passed. They devised 'Proposition 187' that sought to eliminate the provision of social, health care and educational services to undocumented immigrants. This Proposition was approved by 59 per cent of California voters, and became law in November 1994. Yet it immediately came into conflict with US Federal laws, especially the stipulation that denied the children of undocumented parents the right to education. Ultimately, after a long political battle, the Federal government deemed it unconstitutional five years later and the California governor eventually abandoned idea of the Proposition (Alvarez and Butterfield, 1997; *The Guardian*, 30 July 1999). Although this Proposition failed, it did show the capacity of sub-national bodies, in this case an American state, to regulate the contours of citizenship and 'belonging'.

The role of non-state agencies

In the interest of brevity, we shall limit our discussion here to international migration. There are at least three principal types of private (non-state) actors involved in the regulation of international migration. The first is private carriers and related institutions – such as airports, airlines, and ferry services. The second involves formal intermediaries – that is labour brokers who act as recruiters for employers in the receiving countries, and the third involves a set of more clandestine intermediaries – in other words traffickers and smugglers.

With respect to private carriers, the European Union serves once again as an illustrative example, but in this case, non-state forces are reinforcing state regulation. The gradual privatization of immigration control has its roots in the 1990 Dublin Convention, with its policy of 'confinement' for the EU. The policy of 'Confinement' aims to limit the number of potential asylum seekers by distinguishing between 'true political refugees' and 'economic migrants'. This was to be achieved by implementing visas from countries where there were potentially a large number of asylum seekers, and by fining carriers who transport passengers without the proper documents (so-called carrier sanctions). In most European countries, airlines and other transport services can be fined about US$3000. The fines are normally levied on carriers unless an undocumented passenger either requests asylum, is admitted, or when the carrier can demonstrate that the documents presented at the time of departure do not show any 'manifest irregularities'. But as the transport companies have no way of determining the 'manifestly unfounded' character of asylum requests, and even less ability to guess the decision of border officials, it is in their interest to refuse any passenger who does not have the correct documents. The result is that many potential asylum seekers and thus potential workers do not ever reach the shores of the European Union (Samers, 2003).

The second set of private actors includes legal labour brokers. This is a story not so much about regulation, as it is about recruitment, and it especially concerns employers in the Gulf States and Southeast Asian migrants. For example, migrant workers queue up in Bangladesh's Dhaka airport in the uniform of their employers, having paid a considerable sum to the broker to arrange passports, visas, flights, and accommodation. And while many governments actually set a limit on what the brokers can charge (the Thai Ministry of Labour imposed a US$2,240 maximum fee in 1995), many migrants are reported to pay more (Stalker, 2000), thus blurring the distinction between labour brokers and the more illegal forms of trafficking. This leads to our third group of non-state actors – clandestine intermediaries, in other words smugglers and traffickers. Salt and Stein (1997) aptly called this new phenomenon 'Migration as a business'. It is 'new' precisely because of the increased determination on the part of the advanced economies to regulate low-income/low-skilled immigration and asylum seeking.

MIGRATION AS A GEOGRAPHICAL STRATEGY

In the introduction to this chapter we stated that im/migration was a two-sided phenomenon. For some workers it is a choice, for others a last-ditch option. For our purposes here, we can make a simple heuristic distinction between im/migration as a 'proactive choice' and a 'reactive necessity'. In the first case the workers involved have greater latitude to exercise their agency than in the second case; in both

Source areas

Regular employment

Quality of employment (pay, conditions)

Family attachments

Freedom to express one's beliefs and values

Attachment to place

?

Destination areas

Avaliability of employment

Social networks (family contacts)

Freedom to express one's beliefs and values

Knowledge of language/culture

Imagination/perception of destination area

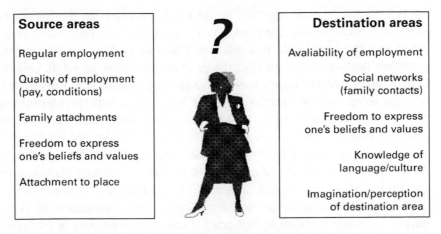

Figure 7.2 Should I stay or should I go? Migrant decision making

cases, though, im/migration can be seen as a geographical strategy to improve workers' fortunes. The choice of migrating however, is usually never an easy one, largely because the migrant has to balance what might be left behind in the source area with what s/he might gain in the destination area. In Figure 7.2, we illustrate some of the reasons for staying or migrating.

Reactive necessity

In situations where migration is 'reactive' (that is, forced, coerced or otherwise undesired but necessary) it is very much a geographical strategy of last resort. As the economist J.K. Galbraith once put it, 'migration is the oldest action against poverty'. Indeed, migration can be a means to escaping unemployment or underemployment (that is working for extremely low wages). True, migrants may find that they are working for extremely low wages compared to native workers in the country of destination, but often what they earn is considerably more than what they could earn in their home country. For example if one examines the sectors in which Indonesian immigrants typically take jobs in Malaysia, they could earn US$2 a day in Malaysia in 1997, compared with only US$0.28 in Indonesia (Stalker, 2000: 22).

Yet we must not assume that migrant workers simply shop for the highest wages. If this were true, then migrants from the poorest countries would move to the richest, but this is not necessarily the case. Often the job search can be geographically limited to neighbouring countries (Sassen, 1995) and im/migrants will simply seek out better pay and not necessarily the highest pay they could earn. For example, Sassen writes about a Mexican immigrant working in a factory in New Jersey who communicates the wages paid to a relative. The relative does not necessarily compare his/her skills, qualifications, and capacities to what can be earned elsewhere, but rather sees it as simply an economic improvement. In other words, the decision to e/migrate is complex and dependent upon the specific circumstances both in the countries (and localities) of origin and destination. Indeed, people e/migrate for a 'better' life – whatever 'better' might mean. Sometimes this has nothing to do with economic issues, but instead means being closer to a relative. Yet, that im/migrants can find a 'better life' should suggest to us that their status as 'outsiders', 'non-locals' or 'ethnic minorities' is not necessarily a hindrance to occupational or wage mobility. In fact, one's status as an immigrant or one's particular ethnicity may actually be *an advantage* in a person's ability to find employment and move up the occupational ladder.

In any case, a former British government minister, Norman Tebbit, once (in)famously said that people who live in areas of high unemployment should 'get on their bikes' and seek jobs elsewhere. His glib statement suggests that migration is a painless process. But it underestimates the psychological stress many workers and their families experience in having to look for work elsewhere – in their own country or abroad. It also ignores the often disastrous effects out-migration can have on the source communities, whose economic and social plight may be exacerbated by the loss of many home-grown men, women and children. Let us look at two case studies, one which documents forced international migration, and the other that examines forced internal migration.

From Casa Blanca to Tulsa The town of Casa Blanca in the Mexican state of Zacatecas provides a dramatic case of how forced migration to the United States simply reinforces the imperative of migration. The majority of the 3,300 people who left Casa Blanca during the 1990s moved to Tulsa Oklahoma, largely because of declining Mexican government agricultural subsidies in a region plagued by drought. In fact, the money which migrants remit to Zacatecas (an estimated US$1 million each day!) is far more than what the Mexican government currently gives to this region in a year. The irony for the residents of Casa Blanca is that as more and more migrants leave for the United States (and remain there because of tightened surveillance of the US–Mexican border) in order to search for work to feed their families, the town becomes increasingly depopulated. This then forces shops and other services to cease their business, which in turn forces others to migrate to the United States. The end result is that large swathes of central Mexico are facing depopulation and economic stagnation (*New York Times*, 17 June 2001). In other words, people living in Casa Blanca are forced to migrate in order to survive – it is a reactive necessity.

From Tulare to Little Rock Some workers are actually paid by the state of California to leave California. Let us consider the case of one woman from the town of Ivanhoe, trying to make a living in the otherwise fertile agricultural region called the 'Central Valley'. After cleaning budget hotel rooms, working for housebound elderly people, picking vegetables, or most recently, surviving on a US$616 a month welfare cheque, she decided that 'enough was enough' in a county in which unemployment levels have hovered between 15 and 20 per cent. Broke and owing three months' rent on her US$300 a month apartment, she decided to pursue the extraordinary option of being paid by the Tulare County government to leave California. From 1998 to 2001, the Tulare government has paid over 750 welfare recipients to leave the county. In June 2001, the woman and her two daughters left California for Little Rock, the capital of Arkansas, armed with US$3,000 worth of Tulare County cheques. She did not have a job lined up in Little Rock, but she had social connections there, and was likely to find work in a city where job vacancies were ample (*New York Times*, 18 June 2001). But it would be too simple to equate such internal migration with high unemployment in one area and low unemployment (and possibly higher wages) in another. Since labour has to be 'socially reproduced' (see Chapter 2), the presence of affordable housing, 'quality' schools, cheap public transport, and other facilities (religious, health-related, etc.) all constrain or facilitate labour migration. In contrast to Mexican migrants leaving Casa Blanca, in the above example, survival was difficult, but not impossible, and the advantages of moving to Arkansas had to be weighed very carefully. Thus, migration was 'forced' but not in the same way as it was for Mexicans.

Proactive choice

So much for labour migrations that are compelled rather than voluntary. The geographical movement of workers need not always be a reactive response to their local circumstances. In other cases, labour migration can be regarded as a positive, proactive strategy by workers to improve their wages, their skills, their employment prospects or their general quality of life. This brings us to the issues of agency and scale. In the previous section we discussed cases where workers were forced to migrate and in which they could not be said meaningfully to have 'constructed' new geographical scales. However, in other cases the connection between worker agency and the making of new scales is an important and reciprocal one.

We can illustrate this with two case studies, one of professional workers in relatively high-paid sectors (that is the financial sector) and one of less well-paid professional workers taking advantage of employment opportunities within their firm.

In the financial sector Professional labour migration in financial sectors in key cities has expanded dramatically in the last 20 years. The reasons for this include the growth of (especially foreign) advanced producer services (that is high-level banking, accounting, law, and consulting services), the internationalization of the financial system, the establishment of international financial institutions, and the emergence of more precarious forms of professional employment (fixed-term projects, etc.). For example, in the investment banking and accounting sectors in certain 'global cities' (for a discussion of global cities, see Sassen, 2000) such as Hong Kong, London, Los Angeles, New York, Singapore and Tokyo during the 1990s, the number of professional overseas temporary secondments increased significantly. These typically had a duration of three years and involved primarily men moving. In some banks, the number of professionals abroad ranged from approximately 100 to 600 individuals. But in many cases, this is not an on-demand labour migration, but actively organized in advance. For example, some Japanese banks had a complex programme of sending workers overseas for international experience, dividing workers into different categories such as 'international expatriates', 'international assignments', and 'short-term assignments' (Beaverstock and Smith, 1996; Beaverstock and Boardwell, 2000). Why do executives and other managers migrate? Among the most important determinants is the accumulation of international experience, which is seen as central to career development, the expectation of high salaries in select cities, and the appeal of a certain set of global cities. As Beaverstock and Smith (1996: 1381–2) highlight:

> The magnetism of London as a working and living space must not be underestimated in the construction and uneven distribution of skilled international labour migration in the world economy's financial industries. Global cities like London are where the corporate professionals want to work, to consume, and to accumulate wealth.

In any case, our point is that this type of labour migration is far from forced. It is a voluntary movement by professionals to improve their long-term career prospects, and has little to do with the other kinds of forced migrations we have talked about.

From Ebbw Vale to Ijmuiden: steel workers on the move The town of Ebbw Vale in South Wales has been making steel for over two centuries. In July 2002 the last major metal plant in the town closed down, marking the end of a way of life. Faced with foreign competition, South Wales steel factories have closed in quick succession since the early 1980s. The results have been unfortunate: because towns like Ebbw Vale have been so dependent on one major economic activity, plant closures have left high levels of local unemployment in their wake. It was in this context that the Anglo-Dutch firm Corus made 500 Ebbw Vale steelworkers redundant in mid-2002. Since many of these workers were among the best paid in the town as a whole – earning between 25 and 30 thousand pounds per annum – the factory closure was a major economic blow. The redundancies at the plant were among 6,000 made in Corus plants worldwide since 1998, half of them occurring in South Wales alone. Part of the reason for this geographic concentration of job losses was the strong British pound, which made steel exports from the UK very expensive for foreign buyers. As a result, over half of the metal making machinery from Corus's Ebbw Vale plant was physically removed to the Dutch steel town on Ijmuiden in Holland (this epic act indicates what is involved in 'disembedding' a firm from a particular place). The euro, being a weaker currency than the British pound, made the relocation economically viable. Put differently, without the geographic transfer of extra capacity, the Ijmuiden plant and its 9,500 strong workforce could not meet the European and worldwide demand for Corus metals.

Though the closure of the Corus plant further accelerated Ebbw Vale's downward economic spiral, it was not a disaster for everyone. In 2002 Corus offered around 100 of its Welsh workers the opportunity to voluntarily move to other Corus plants in the UK and Europe, including the Ijmuiden plant in Holland. Though the cost of living in Holland remains lower than in the UK, Corus not only guaranteed Ebbw Vale migrant workers their same salary (a salary higher than most equivalent Dutch Corus workers), but subsidized house purchases for these workers in Ijmuiden (in effect offering them a 'company mortgage'). Accordingly, many Ebbw Vale Corus workers voluntarily relocated themselves and their families to Ijmuiden in 2002.

In many ways, this does not seem to be a good example of migration as 'proactive' worker agency. After all, the workers involved lost their Welsh jobs; they had to move from a town where most of them had been born and raised; and they moved to a country where the language was foreign to them. This hardly sounds like a convincing example of voluntary migration. However, despite appearances, the workers involved did not simply see it as a 'negative' move that had been forced upon them. In the first place, the economic decline of Ebbw Vale had, in their eyes, ceased to make it a desirable place in which to live and raise a family. With teenage drug use, crime and unemployment all serious local problems, there were few well-paying jobs left in the town once the Corus plant closed. As one of the migrant workers said of Ijmuiden, 'It's an opportunity for the children when they grow up. It's frightening to think of them growing up in [Ebbw Vale] ... there's nothing here for them' (quoted in Gow, 2002: 26). Aside from being a relatively prosperous town compared to Ebbw Vale, Ijmuiden had other attractions too. Specifically, Corus faced a skills shortage in its plant there. With 77 per cent of Dutch people working in the service sector, the engineering skills of the Ebbw Vale steelworkers were in demand. Furthermore, most Dutch speak English well because of their schooling, easing the cultural challenges faced by the Welsh workers in leaving the UK. As one Ebbw Vale out-migrant and his wife commented, 'We worked out all the pluses and minuses on a piece of paper. There just didn't seem that much against coming out [to Ijmuiden].' (For more on this case study see Gow, 2002.) If we link all this back to the general discussion of worker agency in Chapter 6, in this case the structural conditions confronted by the 100 Welsh Corus workers provided more opportunities than they did constraints.

THE MANY SIDES OF MIGRATION

As this chapter has shown, labour migration has social consequences for many different working and non-working groups at a variety of spatial scales. This indicates that any assessment of labour migration must be multi-perspectival. It must, that is, be grounded in the diverse needs and wants of the individuals and groups of workers who are implicated in migration. Thus, migration will have an impact not only on native workers, but also already settled immigrants, other recently arrived migrants, the unemployed, workers remaining in the source areas, and the local and national economies at both ends of the migration stream. For example, the effects of undocumented Mexican immigration on, let us say, low-paid African-American workers in a particular area or industrial sector within Los Angeles, will be different from the aggregate effects of *all* undocumented immigration on the entire American economy. The latter has arguably implications for *all* African-Americans in the United States, as well as on workers in the source areas in Mexico.

Let us begin by looking at the *national* scale, and taking the example of the United States. While immigrants in the United States seem to increase the earnings and occupational mobility of some natives, particularly business owners and others who own a substantial proportion of American assets, it seems to do very little for, or even harm the bulk of native workers. Moreover, those who benefit financially seem to be mostly 'white' and highly skilled, and those who lose in terms of wages, mostly 'black' and less-skilled (Sassen, 1995). Nonetheless, as Borjas (1999) writes in the context of his research on the United States, one can '...conclude – by picking the 'right' period, the 'right' group, and the 'right' methodology – that immigration has either a hugely beneficial or a very harmful impact on the labor market opportunities of native workers' (p. 70). Yet labour market opportunities are not the only way of gauging the impacts of migration. Indeed, both native *and* immigrant workers may benefit because of the *fiscal* impacts of migration (that is migration may increase tax revenues at the local or national scales), which in turn may improve public services or infrastructure for everybody. This should simply re-affirm that scale is central to any understanding of how migration affects different sets of workers.

In that sense, let us now focus on the *local* impacts of migration. Why is it necessary to do so in the context of migration? To begin with, it is important to remember that immigrants in at least the advanced economies tend to concentrate in large metropolitan regions – what William Frey (1996) calls 'gateway cities' in the context of the United States. This does not mean that they are necessarily restricted in settling elsewhere, nor that they do not settle in smaller towns and rural areas throughout the world. Witness the concentration of Mexican migrant workers in the agricultural town of Watsonville, California, Moroccan workers in Roubaix in Northeast France, or Pakistani migrants in Bradford, England. Nonetheless, they tend to settle first, and remain in large metropolitan areas. One possible result of this is that the concentration of immigrants in large cities and their disproportionate over-representation in low wage jobs (notwithstanding the growing number of highly skilled immigrants) depresses wage levels or hinders the employment prospects of native workers in these cities.

But as we noted above, the evidence for the impact of immigrant workers *at any scale* is far from conclusive, whether this is a matter of job competition, wages, or fiscal effects. That said, at least one common way of evaluating the local impacts of migration on employment is to ask whether migrants act as a *complement* or a *substitute* for native workers. Roy (1997) explains the distinction. Migrants are said to be *complementary* to native workers 'if there are skill shortages in the host country and immigrants relieve these bottlenecks, it would expand job opportunities in general, resulting in an increased demand for labor and eventually leading to higher wages for native-born workers' (p. 152). Furthermore, it is often the case the migrants perform jobs which natives cannot or do not wish to do. Alternatively, immigrants and native-born are *substitutable* 'if variations in the number of immigrants relative to the native-born workers across the selected local labor markets demonstrate that a high ratio of foreign-born to native-born workers is associated with a lower wage rate for the native born' (Roy, 1997: 152). An example of this is to be found in Box 7.3.

Box 7.3 Complementarity and substitution in the French automobile industry

In the past, the French automotive industry (and the Paris region in particular) relied on massive numbers of unskilled and semi-skilled immigrant production workers (*ouvriers specialisés* – OS). During the so-called 'Fordist' period (that is from the 1950s to the early 1970s), the number of more skilled workers (usually 'white') (*ouvriers professionals* or OP) were replaced gradually by immigrant OS from Southern Europe. The substitution of the OP by the OS coincided with a period of 'de-skilling', shift-work, and task fragmentation on the assembly lines (so-called 'Taylorism'). Factories became dirty, noisy and largely intolerable places for native workers. Yet as the southern European economies expanded, many of these immigrants gradually returned home, and Moroccans and Algerians were recruited in the late 1960s and the early 1970s to replace them. However, the economic downturn in the advanced economies in the mid-1970s onwards, combined with increasing Japanese competition (in the form of more efficient and higher quality production systems – so-called 'Toyotaism') forced the French automobile producers to respond. As Toyotaism progressed in the *French* factories, immigrant workers especially, lost their jobs as their posts were being gradually eliminated. Thus, from the 1950s until the mid-1970s, immigrant workers acted as a *complement* to native OP, but as Toyotaism progressed from the 1980s onwards, low-skilled immigrant workers found themselves *in competition* increasingly with more highly-skilled production workers (once again mostly 'white'). Today, the automobile plants still rely on a certain 'window' of less-skilled (predominantly migrant) labour to perform tasks which cannot be profitably automated and/or do not require more advanced skills. Thus, while immigrants still remain in competition with white workers for skilled jobs in the automobile plants, those who remain in employment actually *complement* the more skilled native workers (for more on this, see Samers, 1998).

Yet the impacts can be *trans-local* as well. For example, growers in Baja California (in Mexico) cannot compete in terms of wages with growers in California in the United States (who also use immigrant labour, but who siphon off generally the most productive workers). In order for Baja growers to attract workers (they have difficulty retaining them because of the minimum wage in California), the Baja growers continually recruit workers from southern Mexico (Oaxaca), and promote settlement in Baja, rather than increasing wages, or rewarding the workers who have the highest productivity. Thus, growers in Baja California can compete with wages 2,000 miles away and attract workers rather than compete with the California market 200 miles away. But it is an integrated and in many ways trans-local labour system because migrants in Baja accept lower wages because of the possibility of moving to California, while Baja wages place pressure on California wages (Sassen, 1995).

Finally, the migration of workers can even have implications for those workers remaining in the source area. Indeed, while immigration to the advanced economies might fuel growth in those regions, it may have the opposite effect on the countries of emigration, but this too is an issue of scale. First, emigration may have a different impact on relatively low and high unemployment countries. For example, mass emigration may reduce unemployment in the source area. However, it may also lead to the flight of highly-skilled/highly-educated workers (so-called brain drains) which might in turn also increase unemployment by reducing the rate of capital accumulation in the country of emigration. Second, financial remittances (in other words, money sent back to the source area by emigrants) may benefit individual families in the source area. Indeed among Mozambican migrants in South Africa, 87 per cent had saved sufficient funds to purchase a house in Mozambique. Or remittances can provide necessary foreign currency reserves to meet external debt requirements. For example, in Lesotho, the remittances from migrants who work in the mines of South Africa account for nearly a tenth of Lesotho's Gross Domestic Product (*The Economist* 2 September 2000). Yet such remittances can also lead to inflation by expanding the supply of money without increasing productivity or the productive capacity of the source area. In other words, the money is spent on luxury items, often imported, rather than on increasing innovation in the national economy. For example in the towns and villages that surround the city of Agadir in southern Morocco, returning migrants have constructed enormous houses, thus stimulating a property boom, and raising the cost of land and other goods and services in the surrounding area.

The result is that many Moroccans who do not emigrate suffer the consequences of inflation (Berriane and Popp, 1997).

CONCLUSIONS

This chapter set out to explore the various dimensions of labour migration. We began with a discussion of the different types of labour migration and the diversity of immigrant workers. We then proceeded to illustrate how the vast numbers of workers on the move are doing so between an extraordinary range of source and destination areas. Next we explained the regulation of labour migration at the macro-regional, national, and local scale. We argued throughout that regulation remained primarily a national matter, even if there are slow movements towards a privatization of immigration control and more macro-regional migration regimes, such as in the EU. The volume and geography of migration would be vastly different indeed without this regulation, and while such regulation occurs at a variety of scales and mediated by a range of social institutions and actors, the national scale (that is the national state) has remained dominant. True, national borders may be porous, but workers are certainly not free to move at will, and such an argument contrasts sharply with the thesis of a 'borderless world'.

Finally, we discussed the implications of different 'types' of migration and their implications for native workers, other already settled immigrant workers, and even those who never emigrate. We argued that the migration of workers is not necessarily 'progressive' or 'regressive'. Some native workers will 'benefit' from migration in the form of a higher paying job and some native workers and already settled immigrants will be left with unemployment. Legal migrants may take jobs from undocumented workers and vice versa. And the impacts are not simply local, migrant workers may expand or hinder the prospects of other workers far from the destination area of migrants. This is because migration has other effects (such as expanding national economies) which are separate from local employment impacts. In short, scale matters! Yet even if the effects of migration are scale-dependent and difficult to measure, what *is* certain is that under capitalism, workers will remain ever restless in the search for a 'better' life.

eight

Up-scaling Worker Action

In the previous two chapters we have presented examples of wage workers using geographical strategies to effect change in the conditions under which they live and labour. However, in many cases the kind of place-based action discussed in Chapter 6 is insufficient, while migration, as set out in Chapter 7, is neither possible nor desirable. This raises a question: can workers remain in place and yet act translocally in order to further their local needs and wants (or those of others elsewhere)? As this chapter will show, the answer is emphatically 'yes'. If the case studies in Chapter 6 demonstrated that workers often 'think locally and act locally', those presented in this chapter show that they can also (i) 'think locally and act translocally' and (ii) 'think *and* act translocally'. In other words, workers and other social groups sympathetic to their cause can 'up-scale' their struggles and pursue them on a broader socio-geographic terrain without resorting to migration – just as the Tompkins Square Park homeless did in Smith's discussion of what he calls 'scale politics'. These up-scaling efforts can be undertaken by and on behalf of workers in one place, or many workers in several places. Unlike the examples discussed in Chapter 7, the attempts to transcend the confines of the local discussed in the pages to come involve the seeming paradox of remaining place-*based* while acting *across* space. But we shall see that this is not a paradox at all:

for in our interdependent capitalist world, the defence of local livelihoods often necessitates reaching out over broader geographical scales precisely because, as Massey (1993) argues, these days the translocal is *in* the local. Put differently, contemporary workers need a 'global sense of place'. In Swyngedouw's (1997: 168) words, they increasingly need to 'reach ... out from local ... identities to find threads that enable solidarity and extend lines of power for those that remain otherwise trapped in place'. In situations where purely local efforts to defend jobs are ineffective and where out-migration is not viable, workers and others must actively forge translocal ties in order to realize local interests. In effect, this involves labour 'matching' the translocal organization of many firms and regulatory institutions. Without this effort of up-scaling, the idea of 'hegemonic despotism' discussed in Chapter 5 would probably ring more true: for firms and many regulatory bodies would be able to 'play off' workers in different localities against one another more readily than they already do, in what's variously called 'whipsawing', 'the race to the bottom' and 'geographical divide-and-rule'.

The up-scaling of worker action – involving workers acting alone or in concert with others – is evidence of what we earlier called a geographical imagination. It is one that lies at the heart of the modern labour movement. Such an imagination, to quote Harvey's (1973: 24) germinal definition, 'enables ... individuals[s] to recognize [how they] are affected by the space that separates them, ... to judge the relevance of events in other places, [and] to fashion and use space creatively'. Until relatively recently, the 'space' that Harvey refers to was not necessarily global in scale. It was often national, or at best macro-regional, in scale and trade unions were the principal worker institution through which the up-scaling of worker action against firms and regulatory bodies was organized. Today, however, we have seen that workers in one place must recognize that their fate is deeply dependent on the actions of myriad workers in far-flung corners of the globe. Not only is the 'working class' far larger numerically than at any point in human history. It also encompasses virtually every place on the planet, not to mention every creed, ethnicity and colour. Trade unionists have responded to this fact by creating new, or strengthening existing, *international* and *global* unions (introduced in Chapter 4). However, their undoubted importance notwithstanding, these unions are neither necessary nor sufficient for the up-scaling of labour struggle. Workers in the less developed world, the former communist bloc, and newly industrializing places often lack union representation. At the same time, the 'workerist' history of the union movement means that it has often failed to link worker struggles with those of others in civil society. Accordingly, the contemporary landscape of transnational worker action extends beyond union organizing to encompass a range of non-union initiatives. This combination of union and non-union action is equipping contemporary workers with the means to overcome the vulnerabilities of being rooted in place.

What are the potential benefits of inter-place cooperation between workers and those with worker interests at heart? Three are immediately obvious. First, though workers remain rooted in place (migrants excepted), connecting across space provides an obvious mechanism for counteracting the kinds of sorry tales recounted in Chapter 5. Put simply, inter-place cooperation can prevent workers being played-off against each other. Second, there is strength in numbers. The larger the scale of labour organizing, the more people can be deployed to challenge those firms and regulatory institutions who might be threatening workers' livelihoods. Finally, there are logistical advantages. Unions and other organizations operating at the national and transnational scales usually have considerable resources at their disposal by virtue of their large memberships. These resources, in turn, can underpin major initiatives (strikes, for example). In short, there have long been (and remain) compelling reasons for wage workers and allied groups to up-scale their actions beyond place.

The chapter is organized as follows. In the next section we look at national trade unions, the 'classic' geographical scale and institutional means at/through which workers have up-scaled action from the local level. Following this, we move to the international and global planes, and deal with worker actions in and around commodity production and commodity consumption respectively. The latter is a relatively new and increasingly effective form of transnational worker action, and is linked to ideas of 'ethical consumption'. It is also a form of worker action that is often separate from formal trade union organizing. The penultimate section takes this non-union theme forward to consider so-called 'new social movement' activism at the transnational scale. This form of activism links worker struggles to those of groups who do not define themselves or their interests primarily in class or occupational terms. It is a form of activism equally at home in formal union settings and looser non-union contexts. The chapter ends by evaluating the up-scaling of worker struggle. Mirroring the discussion of local action at the end of Chapter 6, we argue that it is naïve to regard translocal action by workers as intrinsically 'progressive'. Though many labour activists and Left-wing labour analysts are generally 'for' up-scaling struggle and 'against' exclusively local level worker action, we argue that even when workers reach out and construct new scales (or new relations within existing scales) the geographical dilemmas highlighted in Chapter 4 arise. We explore whether or not these dilemmas are resolvable in the next and final chapter.

TRANSCENDING THE LOCAL: NATIONAL UNIONISM

Turning outwards as a form of defence (or attack) is a venerable geographical strategy among wage-workers and their representative institutions. Trade unions have historically been the principal bodies that agitate for workers' interests within and beyond the workplace (see Box 2.3). They are, to quote the Webbs' (1920: 1) classic definition 'a continuous association of wage-earners for the purpose of maintaining or improving the conditions of their working lives'. These bodies, until recently, have been predominantly national in their structure, being composed of myriad local memberships. Operating under a national leader and an executive, these memberships (at least in democratic nation-states) both support their union financially and influence national decisions through ballots and other means. National unions began to flourish in the Western industrializing countries in the nineteenth century, since when they have been established in many other countries worldwide.

Why, it may be asked, did *national* unions historically come to prominence as opposed to sub-national or global ones? And why do these unions remain important for workers today? We argued earlier in this book that place is the most obvious scale at which workers develop loyalties to one another by virtue of their daily contact and personal investment in a given location. This implies that any effort to up-scale worker solidarity requires some means for getting geographically separated employees to identify common bonds. These common bonds, historically, have been provided by national identities. As political theorist Benedict Anderson (1983) famously argued, nations are 'imagined communities' that are manufactured through the mass media, key 'opinion shapers' (like politicians) and public institutions (like schools). Thus, there is nothing natural about 'Britishness' or 'the Brazilian character' for example. These are fictions that nonetheless have real effects insofar as people in Britain and Brazil internalize these labels and what they supposedly stand for. Because of the power of ideas of national character, and because nations have become the major unit of political organization in the modern world, it was perhaps inevitable that workers would initially organize across space at this particular scale.

What this means is that for many decades wage workers in countries with strong national union traditions have defined themselves in national (and sometimes *nationalist*) terms as much as in local terms. Thus they think of themselves as *Canadian* workers or *Uruguyan* workers, rather than as workers in general or (say) *Torontonian* or *Montevidean* workers in particular. In countries where national unions are able to exert considerable influence over firms and regulatory institutions they have a number of key powers at their disposal. These powers are physical (for example, the capacity to call national strikes), financial (for example, the capacity to fund expensive advertising campaigns to change public opinion in their favour), legal (for example, the capacity to use in-house lawyers to defend worker rights in courts) and political (for example, the capacity to influence government policy by lobbying politicians, to produce policy documents on key topics like 'living wages', etc.). In each case, these powers derive from numerical strength and 'critical mass'. That is, they derive from the capacity a union has to speak for and to mobilize workers across a whole national territory. Where, in the case of very large national unions, the membership numbers millions then it is plain to see that these unions can be powerful institutions. This power can be exercised to support just one worker or one workforce in one place; or it can, at the other extreme, be used to further the interests of the national membership as a whole.

It is important to understand that national unions both past and present are *scale-constructing institutions*. They do not simply 'fill' an existing national scale like water in a glass. Rather, they add new sets of relations and capacities to the national landscape of labour in any given country. There are too many examples of national unions successfully aiding some or all of their members at the national scale for us to consider here. The one we do focus on has the virtue of showing a union actively making scale and, in the process, enabling its members to achieve things they could not achieve by acting alone at the local level. It's an example drawn from Herod's (2001) landmark book *Labor Geographies* (see also Herod, 1997).

During the 1950s, US dock workers did not have effective national union representation – just as so many workers today do not, especially in the developing world. Most US ports then, as now, are situated on the east coast (stretching from Maine to Texas). Though east coast dockers had a national union in name during the 1950s – the International Longshoremen's Association – in practice it did little to unite these dockers in any meaningful sense. Symptomatic of this was the fact that negotiations over dockers' wages, hours, rights and conditions were conducted on a port-by-port basis. New York dockers, working in the east coasts' largest port, enjoyed the best of things, while those in smaller ports did relatively less well. If a shipping company, say, considered port charges to be too high, then it could (in theory at least) choose a cheaper east coast port to traffic goods through.

Until the late 1950s US docks were major employers of wage workers. Loading and unloading ships was a highly labour intensive activity, utilizing tens of thousands of dockers. But a technical innovation changed this virtually overnight in the US and elsewhere: namely, containerization. By packing goods into large, standard sized metal containers, it became possible to use machines to load and unload ships more than had previously been the case. Thus, where the port of New York employed some 80,000 workers in the 1950s, today it utilizes just a few hundred (Herod, 1997). Containerization thus posed a serious threat to dockers: directly, by displacing manual labour, and indirectly by lowering the labour costs and increasing the un/loading speed in ports where containerization was first embraced. These indirect effects meant that, in the 1950s, 1960s and 1970s, ports still using a lot of dock labour could ratchet down wages because of competition from other ports where containerization was favoured.

In light of this, the ILA determined to nip the problem in the bud by pushing for *national* salaries and *national* standards for east coast dockers from the late 1950s onwards. This was no mean achievement. In the first place, which salaries and which standard should become the norm for all, those in New York or elsewhere? Second, the ILA faced competition from a rival southern union, the International Brotherhood of Longshoremen (ILB). The ILB feared that the ILA was too New York based and not representative of all US dockers' interests. In particular, southern US dockers faced less of a threat from containerization because they handled more agricultural products than ports like New York. These products were less amenable to container storage. Finally, the ILA pushed for a national agreement with dock employers at the very time when containerization made dockers less of a strike threat because there were fewer of them.

Despite all these challenges, the ILA made real headway during the 1960s. First, it used persuasion at the grass-roots level to show dockers that they all faced a common threat from containerization – if not immediately then in the future. Second, it remained flexible about which national salaries and standards it should aim for. Aiming too high might make employing dock workers too costly, and exacerbate unemployment. Consequently, the ILA argued for a national 'master contract' for all full-time dockers, with 'local contracts' adjusted to suit circumstances so long as the master contract was not flouted. On this basis the ILA was able to mobilize most east coast workers to go on strike – or to threaten to – several times during the 1960s. The result, after much lobbying of dock companies, was a 1971 east-coast wide employment agreement, which was improved upon in 1977. A 20-year struggle to create some geographical uniformity of pay and conditions was thus achieved, albeit in the context of increased unemployment as containerization took hold. As Herod (1997: 146) puts it, 'the ILA ... sought nothing less than to construct a new geographical scale of bargaining and labor relations in the [dock] industry'. A set of rules and procedures that previously had not existed were literally made by the ILA in negotiation with employers and regulatory bodies. These rules and procedures then took on a life of their own, as it were, and undercut the divide-and-rule history of port-by-port bargaining in the US. Though a historical example, the ILA case stands as an inspiration for, and a challenge to, those workers who currently lack national union representation. Countries where such representation has virtually never existed will have a long, slow job of making national unions a reality; in countries where national unions have been beaten-back by aggressive governments and employers (like the UK), these unions must rebuild. Despite the undoubted difficulties, the ILA case shows what is possible with effort and determination.

A NEW WORKER INTERNATIONALISM: BORDERLESS SOLIDARITY

National trade unions and their local memberships, as the case study above showed, still play an important role in contemporary labour struggles (see also Rutherford and Gertler, 2002). However, as the twentieth century has given way to the twenty-first, there has been a progressive 'de-coupling' of the geographical scale of economic activities and the scale of union organizing. In other words, it has become increasingly apparent that national trade unions are not sufficient, on their own, to address the transnational processes that impact on the local lives of wage workers, their dependents and other civil society groups. In order to match the scale of these processes, several existing and some new worker institutions have defined themselves in explicitly non-national terms. The existing ones include International Trade Federations (which are industry based), macro-regional unions (like the Organization of African Trade Union Unity), global unions (for example, the WFTU), and transnational non-union bodies (like the

Box 8.1 Recent activities in transnational labour organizing

Aside from some of the campaigns described in this chapter, a plethora of recent strategy meeting and conferences attest to the excitement among unions and pro-worker groups about the possibilities for transnational organizing. These include:

- A 1998 'pre-millennial' event in 1998 sponsored by the Danish General Workers Union, *A New Global Agenda – Visions and Strategies for the 21st Century*
- A *World Meeting Against Globalization and Neoliberalism* held in Brazil in 1999 sponsored by a range of national trade unions
- An *Open World Conference in Defence of Trade Union Independence and Democratic Rights* held in San Francisco in 2000
- A *Unions and the Global Economy* meeting, convened by US unions and academics in Milwaukee in 2000
- A *Trade Unionism in the 21st Century* conference organized by the Southern Initiative on Globalization and Trade Union Rights (SIGTUR) in Johannesburg in 1999
- A *Building and Labour Movement for Radical Change* conference held in Cologne in 2000 and organized by the Amsterdam-based worker intelligence body the Transnational Information Exchange.
- A *Promoting International Labour Rights and Solidarity* workshop held near Boston in 2000 and organized by unions, academics and NGOs from North and South America

Source: Waterman (2001: 330–1)

ILO or Women Working Worldwide). New ones include those in the European Union, like the European Works Council (see Sadler, 2000). As we write, there is a ferment of activity surrounding the possibilities for transnational labour organizing (see Box 8.1).

As explained in Chapter 4 (in the section 'Re/scaling workers and unions'), there is a long but uneven history of wage workers organizing transnationally stretching back to late nineteenth century Europe and the so-called 'First International' (see Box 8.2). Today, the possibilities for such organizing are greater than ever before. In part this is because modern telecommunications make it potentially easier to connect otherwise separated working communities and their struggles (see Lee, 1999). But it is also due to the fact that many firms are now more *geographically vulnerable* than heretofore. That is to say, the complex and transnational nature of their operations means that there are *many* different places where their operations can be

Box 8.2 Borderless solidarity in the 19th and 20th centuries

The revolutionary and analyst of capitalism, Karl Marx, was a key figure in the so-called First International of the late 19th century. The International was a body devoted to building-bridges between workers in different countries (especially in Europe). By the early twentieth century it had been replaced by the Second International. This was established by socialist political parties, the most powerful (after the 1917 Russian Revolution) being the Russian Communist Party. By 1914 the Second International had 12 million members in 22 countries based on 27 socialist parties. Linked with the two Internationals were the International Trade Secretariats (ITSs). Founded in the 1890s, these still exist today. Despite the long history of worker transnationalism, it weakened during the mid-20th century as national identities solidified courtesy of the two world wars and the subsequent Cold War. Thus *national* trade unions tended to become the principal translocal bodies to which wage workers in many countries pledged their allegiance. For instance, from 1945 very few US workers would identify themselves with Russian workers because of the political gulf between liberal-democratic America and communist USSR. Even at the international level, union organizations splintered along ideological lines, with the ICFTU and the WFTU agreeing to disagree on basic principles. In light of this, the recent attempts to build transnational ties among wage workers within and between the 'first' 'second' and 'third' worlds is very much a renewal and extension of the earlier efforts of the two Internationals and the ITSs. They should not be seen as wholly new attempts. See Logue (1980).

interrupted. We saw an example of this in Chapter 6 vis-à-vis General Motors. If we extrapolate from this case, it is plain to see that workers can potentially conduct a *multi-sited* campaign against a firm and/or a regulatory body across *several scales* in order to pursue their cause. In simple terms, we can distinguish the sites of commodity supply (the places where a firm's inputs come from), commodity production (the places where a firm produces goods and services), commodity distribution (the communications arteries through which commodities move to intermediate and final markets) and commodity consumption (the places of final demand and sale). Depending on the commodity in question, these sites are linked in intricate ways over greater or lesser distances (see, for example, Hughes (2000). The second of them is, if you will, the 'classic' focus for transnational worker actions. But, in recent years, the other three have assumed increasing importance as sites of transnational worker struggle. In light of this, we now consider an example of production focussed and non-production focussed transnational workerism respectively.

The needs and rights of distant strangers: production politics

Transnational worker actions focussed on a firm's production facilities can be undertaken with or without the formal assistance of trade union bodies. The advantages of having this assistance are obvious. But this is not to say that a lack of union involvement is inimical to the success of multi-sited worker actions that cross national boundaries. A good example of such non-union internationalism that is the production-focussed Fair Trade initiatives. These are typically led by a coalition of NGOs and charities. They aim to ensure that Western firms who source products from the developing world do so in ways that are non-exploitative. For example, Café Direct – a Fair Trade scheme that includes big retailers in countries like the UK and small coffee bean suppliers in countries like Colombia – guarantees a minimum coffee purchase price for the suppliers. This prevents the price being beaten down by ruthless retailers and ensures that suppliers can look forward to a predictable annual income (see Whatmore and Thorne, 1997). Important though this and other Fair Trade initiatives have been, such production-oriented forms of transnationalism tend to be more successful when the weight of trade union bodies is behind them. Accordingly, we now turn to a recent example of this.

In 1989 an aluminium plant near Ravenswood, West Virginia, was purchased by a company formed by investors for the purposes of buying the plant (the Ravenswood Aluminium Corporation – the RAC). Almost immediately the Corporation laid off 100 workers, upped the pace of work in the plant and dissolved a joint worker-manager safety committee. Over the next 18 months five workers were killed and several injured in the Ravenswood plant. Most shop floor workers belonged to the United Steelworkers of America (USWA), a national metalworkers' union that was a powerful force in post-World War Two economic life in America. When, in 1990, the Ravenswood branch of the USWA (local 5,668) came to renegotiate its members' contracts with the RAC, it found that 1,700 non-union replacement workers were hired. The RAC's explanation was that local 5,668 had dragged its heels and was, in effect, on strike by not resolving contract negotiations on time. Legally, this argument was sound (courtesy of a Supreme Court ruling). Dismayed and angry, the local workforce decided to fight back through the USWA and the 'general' US trade union the American Federation of Labor-Congress of

Industrial Organizations (AFL-CIO). The AFL-CIO became involved largely because it was alarmed at the use of non-union replacement workers across US firms in the 1980s and 1990s. It thus decided to take a stand with the Ravenswood case.

The starting point for the campaign was an anonymously sent RAC audit that had been undertaken by the accounting firm Price Waterhouse. This audit identified who the owners and the shareholders of the newly formed RAC were. It turned out that the Ravenswood plant was implicated in a global network of investors that the local 5,668 were largely unaware of. Chief among these investors was Marc Rich, a notorious financier who was wanted in the US on tax fraud charges and lived in Switzerland. Additionally, a Swiss finance house, a US conglomerate (Clarendon Ltd) and a Dutch bank had also been heavily involved in the purchase of the Ravenswood plant by the RAC. Equipped with the audit information, the International Department of the USWA and the AFL-CIO organized a multi-pronged campaign to get the Ravenswood workers reinstated. The campaign had an avowedly transnational dimension to it. First, over 300 end-users of RAC metals worldwide were pressured to stop buying Ravenswood products. This pressure entailed verbal and written lobbying by the two unions involved. Notably, a House of Representatives subcommittee was asked to investigate why the US mint had bought US$25 million of metal from Clarendon for making coins. Second, efforts were made to put the spotlight on Rich as the key player behind the Ravenswood drama. In a number of actions involving union members in Western Europe, attempts were made to disrupt economic activities that Rich had a stake in. For instance, in conjunction with the International Union of Food and Allied Workers, 20,000 largely Romanian unionists protested against Rich's attempted purchase of a luxury hotel in Bucharest. By the spring of 1992 anti-Rich activities had been organized in some 28 countries on five continents. These included activities against the Swiss finance house and Dutch bank mentioned above, to which Rich had financial ties (see Table 8.1)

In sum, transnational actions were a key part of a multi-scalar campaign that linked Ravenswood workers with myriad others worldwide. By June 1992 the RAC was forced to concede defeat and to reinstate the fired members of local 5,668. Success, in large part, was achieved by coordinated actions focussed on different parts of the global economic empire that Rich and his associates controlled. The importance of the US unions in the Ravenswood case was to have the resources and the legitimacy to approach unions and workers in other countries. The Bucharest rally was just one of several examples of

TABLE 8.1 Key events in the Ravenswood Dispute

25 September 1990	Negotiations for new contract begin. RAC and USWA exchange contract proposals.
31 October 1990	RAC demands acceptance of the company's final offer; USWA counters and proposes extending current contract RAC rejects extension, locks Local 5,668 members out, and brings in replacement workers.
21 December 1990	West Virginia State Board of Review rules dispute a lockout and awards workers unemployment compensation.
11 March 1991	USWA initiates domestic consumer boycott campaign under auspices of AFL-CIO's Strategic Approaches Committee.
28 April 1991	7,000 supporters of Local 5,668 gather in Ravenswood for Workers Memorial Day observance for 5 workers killed on the job at RAC.
23 June 1991	Local 5,668 Negotiating Committee and IUD/AFL-CIO representatives make first trip to Europe to gather international support.
28 August 1991	Strohs Brewing Company notifies USWA it will stop using RAC aluminum.
23 September 1991	USWA and NLRB lawyers present evidence to Administrative Law Judge that RAC illegally replaced Local 5,668 members.
December 1991	Establishment of European Office to coordinate anti-Rich activities.
4 December 1991	US House Government Operations Subcommittee opens hearings on Mint contracts awarded to Clarendon.
21 January 1992	Anheuser-Busch notifies USWA it will stop using RAC aluminum.
28 January 1992	USWA members return to Europe. Demonstration subsequently organized outside Rich's corporate headquarters in Zug, Switzerland.
4 February 1992	Miller Brewing Company notifies USWA it will stop using RAC aluminum.
14 February 1992	International Union of Food and Allied Workers' Associations trade secretariat organizes anti-Rich demonstration in Bucharest.
16–19 March 1992	USWA briefs representatives of the Inter-American Regional Organization (ORIT) of the ICFTU on Rich's Caribbean and South American operations.
29 April 1992	Negotiations begin for new union contract.
27 May 1992	Tentative agreement reached.
29 June 1992	Local 5,668 workers return to work.

Source: USWA (1992), reproduced in Herod (1995)

non-American workers taking direct action to support a small Ravenswood workforce many miles away (see Herod (1995) for details of the Ravenswood case). It is unlikely that this workforce would have been able to undertake a transnational campaign without the considerable union support it received (though see Castree (2000) for a comparable case where unions were not as heavily involved).

Caring at a distance: consumption politics

Though production-focussed campaigns like the one discussed above are now quite common, those at the other end of the commodity spectrum – that is, the consumption end – are fast proving to be equally effective. These campaigns try to undermine the final market for a firm's goods, thus giving it reason to address worker demands at any of the previous points in the production network. Why, it may be asked, do workers and those representing them need to resort to consumption-related actions? Why not stick with production-related actions like those of the Ravenswood steel workers? There are at least two reasons. First, workers and allied social groups living in the places of commodity production may simply lack the power to resist the pressures that firms and regulatory institutions may put them under. In other words, it is only actors located *elsewhere* along the transnational networks in question that can effectively act on their behalf. Second, particularly when dealing with mass consumption commodities (such as shoes or video games), it is possible to target very large numbers of people in a range of places to aid the worker cause. There are, literally, hundreds of millions of consumers of certain commodities. In other words, consumers are a resource and an opportunity for pro-worker struggles. If some or most of these consumers can, in any given case, be encouraged to alter their purchasing behaviour, then this can put real pressure on the relevant firms and/or regulatory institutions. Consumption-focused campaigns tend to take one of two forms: namely, 'bad publicity campaigns', which make consumers aware of what (unsavoury) labour practices go into making a given commodity; and direct consumer boycotts, where consumers are actively encouraged to no longer buy a particular firm's product. As US geographer Elaine Hartwick (2000) argues, consumption-oriented campaigns by and on behalf of workers can take some of the sheen off the image of the commodities and the businesses in question. By revealing the hidden injustices that might lie behind say, the sale of a Nike sports-bag in a clean, air-conditioned shopping mall, these campaigns can have powerful, positive effects for workers.

Box 8.3 What is a sweatshop?

The word sweatshop was originally used in the 19th century to describe a subcontracting system in which the middlemen earned their profit from the margin between the amount they received for a contract and the amount they paid workers with whom they subcontracted. This margin was said to be 'sweated' from the workers because they received minimal wages for excessive hours worked under unsanitary conditions. Today's subcontracting system functions on a global basis. Large clothing companies produce apparel in 160 countries, often with shockingly low wages and horrible working conditions. Apparel workers in Bangladesh earn around 20 US cents an hour, and in the free-trade zones in El Salvador they earn 56 US cents an hour, just to give two examples. The clothing companies then export that apparel to 30 developed countries, such as the United States and Canada. Meanwhile, apparel workers in the developed world are forced to compete against those conditions. A sweatshop is characterized by the systematic violation of one or more fundamental workers' rights that have been codified in international and US law. These rights include the prohibition of child labour, forced or compulsory labour, and discrimination in employment based on any personal characteristic other than the ability to do the job; the right to a safe and healthy work environment that does not expose workers to degrading or dangerous working conditions; freedom of association and the right to organize and bargain collectively. A sweatshop is also characterized by wages that do not permit workers to adequately feed, clothe, and shelter themselves and their families, and hours of work so long that education and a decent family life are out of reach. Sweatshops are often lawless operations in other ways, evading not only wage and hour laws, but also paying no taxes, violating fire and building codes, seeking out and exploiting undocumented immigrants, and operating in the underground economy, hidden from public view.

Source: compiled by UNITE's Research Department and distributed as part of a Campaign information packet; reprinted in Johns and Vural (2000: 1197).

A compelling, recent example of consumption politics in action is the Stop Sweatshops Campaign (SSC). Led by a US trade union – the Union of Needletrades, Industrial and Textile Employers (UNITE) – and a national consumer rights group (the US National Consumers' League – the NCL), the campaign sought to eradicate the use of so-called 'sweatshop labour' in the production of commodities made in, or imported to, the United States (see Box 8.3 for a definition of sweatshop labour). Such labour is especially prevalent in industries where a firms' commodities – for final sale or intermediate production – are sourced from 'middlemen' and require large inputs of un- or semi-skilled labour. A 'classic' sweatshop industry is the garment trade, where large clothes retailers like Gap source their goods from suppliers far and wide.

Indeed, at UNITE's founding convention in 1995 three teenage garment workers from Central America told an audience of 3,000 about the abuses they faced while making clothes that were sold on to Gap. By 1997, determined to eradicate the low pay, long hours and poor working conditions associated with sweatshop production, UNITE and the NCL had got some 60 non-labour organizations to support its campaign. This was achieved by stressing how sweatshop labour was not simply an issue of *economic exploitation* but a wider question of *morality*. That is, the campaign was 'sold' to non-worker groups in civil society as being about the responsibility ordinary American consumers have to peoples elsewhere who are less fortunate than themselves. More particularly, the campaign emphasized the responsibility of *American* firms to non-American workers. By stressing the nationality of the firms who are guilty of perpetuating sweatshop labour, the campaign tried to make US consumers feel empowered to challenge this blemish on America's 'good name'. On this basis the campaign enrolled religious groups, ethic organizations (Jewish, Latino and African-American), student groups, veterans' organizations, civil rights bodies and other non-governmental institutions.

In practical terms, the campaign involved a mixture of negative advertising against Gap and other firms; educational efforts in shops and malls, where shopper and store managers would be asked by SSC activists if they were aware of the sweatshop origins of the clothes on display; and public events (like university rallies) that emphasized how buying Gap and other apparel firms' products inadvertently perpetuated sweatshop labour in Central America and beyond. In addition to these critical interventions, the campaign took positive steps to eradicate sweatshop production. First, it lobbied the US government to pass a Stop Sweatshops Act that would make firms legally liable if their suppliers violated US labour laws; second, UNITE and the NCL joined President Clinton's Apparel Industry Partnership which aimed to set proper labour standards for garment manufacture in and to the US – and to set up an independent body to check that standards were being adhered to. Finally, the campaign argued for sweatshop free zones within the US – such as schools or whole cities. By 1998 several city governments in America, with local people's support, has agreed to prohibit the sale of sweatshop goods in their locality. In all three cases, the SSC's legitimacy derived from the widespread support it received from consumers in places across the whole of the United States.

In sum, the primary feature of the campaign was 'a focus on consumers as agents of change, the downplaying of tradition union tactics, the shift ... to the site of consumption, and the use of a moral agenda intended to appeal to a broadly defined consumer public' (Johns and Vural, 2000: 1210). The campaign continues. Though it has by no means eliminated sweatshop-made commodities being sold in the United States, it has certainly led Gap and other prominent firms to amend their supply practices. These firms cannot afford to have their brand-image tarnished by association with bad labour practices. The disturbing picture of overworked, underpaid workers in the developing world making items for wealthy developed world consumers does not make for sales and profits. This picture is a 'profane illumination' that can change consumers' buying habits almost overnight and thus put commercial pressure on even the biggest of capitalist firms.

BEYOND WORKERS: SOCIAL MOVEMENT POLITICS

The example of transnational consumer politics discussed above is one of several that challenge the idea that trade unions remain the best or only vehicles for facilitating contemporary efforts to build worker solidarity above the national scale. Though a trade union led the SSC, the campaign was not exclusively union based. In many ways, the strengths of transnational trade unions (and, for that matter, worker organizations like ITFs and IWCs) are their weaknesses. First, they presume a definite membership, which can exclude workers who do not 'belong' to the union in question. Second, they can be very top-down and encumbered not just by their own rules and regulations but those of the nation states in which they operate. Finally, they are largely irrelevant in places where there is no tradition of union organizing and where there is no prospect – for whatever reasons – of unions being formed in the near future. Accordingly, a number of commentators (for example, Hyman, 1999; Thorpe, 1999) have argued that these unions should either be reorganized or formally supplemented by non-union institutions.

In Hecksher's (1988: 177) words, what is required is 'a kind of [transnational] unionism that replaces organizational conformity with coordinated diversity'. This looser, more inclusive, more grass-roots way of working has been described by Kim Moody (1997) as 'social movement unionism'. The term refers to alliances between worker groups and so-called 'new social movements' (like environmental organizations or gay and lesbian groups) that campaign on a single issue or for a single cause. These issues or causes, while not about wage work *per se*, may nonetheless impinge on this question. It's a moot point whether social movement unionism represents an adjustment to, or an eclipse of, standard transnational union practices. According to Waterman (2001) it is a more supple, bottom-up way of organizing against firms and regulatory institutions than one normally associates with trade unions. What is more, Waterman continues, social movement unionism actively emphasizes the *synergies* between workers' interests and those of other groups in civil society. We saw an example of these synergies in the Stop Sweatshops Campaign: what started as a labour issue quickly became a more general moral issue of the proper treatment of people. On this basis the campaign enrolled non-worker groups in its cause. Likewise, in Chapter 6 we saw how 'community unionism' can link worker struggles to those of a broader constituency.

A particularly important focus for social movement unionism at the international and global scales is informal labour. Informal labour is that conducted outside of the 'formal economy'. This type of labour, according to the organization Women in Informal Employment Globalizing and Organizing, includes unpaid work in family businesses, paid work where the worker has no fixed employer, and paid work where the employer evades the rules and regulations of normal economic activity (Gallin, 2001: 537). Sweatshop labour is just one example of informal work. This kind of work is immensely important in both the numerical and the income sense (see Table 8.2). That is, informal work is the main type of employment for people in the developing world, and is the key means of earning an income for millions of labourers worldwide. Such work has three key characteristics. First, it is virtually non-unionised. Second, it is typically low paid and physically demanding because it occurs outside the regulatory frameworks of formal business. Finally, it is overwhelmingly women's work (Gallin, 2001). It's a debatable point whether the category of people called 'peasants' should be included under the informal worker label (Bernstein, 2001).

What has all this got to do with wage-workers in the formal economy (the focus of most of this book)? And what has it got to do with transnational labour organizing? To answer the first question, formal and informal labour are directly linked in two ways. First, as

TABLE 8.2 Informal sector employment

Informal Sector Share of	Latin America, Caribbean	Africa	Asia
Total employment excluding agriculture	15%	18%	15–30%
Total employment including agriculture	45%	75%	75–85%
Non-agricultural employment	57%	78%	45–85%
Urban employment	40%	61%	40–60%
Poor employment	50%	NA	NA
New jobs	84%	93%	NA
	Low-income countries	**Middle-income countries**	**High-income countries**
Total employment outside formal sector	80%	40%	15%

Source: Table created by Jacques Charmes (Université de Versailles, Centre d'Economie et d'Ethique pour l'Environnement et le Développement). Reprinted in Gallin (2001)

companies like Nike demonstrate, many firms sub-contract production to businesses that employ informal labourers. To all intents and purposes, then, the 'core' workers employed by a firm like Nike are in the same production network as 'contingent' workers (the informal ones). Second, when firms are allowed to use informal labourers they can drive down the wages and conditions of formal workers making the same products. In a sense, informal workers are 'unfair' competition for formal workers in the same industries. To answer the second question, all of the above has a geographical dimension. Since informal labour is disproportionately concentrated in the developing world, wage workers in the developed world – where formal employment is disproportionately concentrated – have both a benevolent and a 'selfish' interest in helping informal workers organize against exploitative employers.

For these reasons, several transnational unions and some new non-union bodies have, of late, been trying to organize informal workers into new collectives – both on their own and in alliance with formal economy workers. Examples are the Self Employed Women's Association in India and the home-based workers' office of the Australian Textile, Clothing and Footwear Unions. Such organizations have their work cut out accessing informal labourers, let alone mobilizing them *en masse*. These workers are typically dispersed in small businesses far and wide. Predominantly women, they may also be nervous about fighting back against their employers, especially in highly patriarchal societies. These problems notwithstanding, there are now signs that new union and non-union initiatives are beginning to help informal workers improve their living conditions. Given the sheer number of non-unionized and informal workers worldwide, an inclusive transnational worker movement will surely be one of the hallmarks of labour organizing in the years to come.

PROGRESSIVE TRANSLOCALISM?

All the examples of inter-place cooperation discussed in this chapter are cases of scale politics. They are cases where workers and allied social groups have, quite literally, remade the economic landscape by reaching out from place. What are we to make of these scale-constructing activities? In Chapter 6 we observed that many labour analysts on the Left of the political spectrum take a dim view of resolutely local worker action. The flip side of this is that they take a generally positive view of translocal worker organizing. The reason should by now be obvious. Such inter-place cooperation, in these analysts' eyes, can take workers away from geographical competition and uneven development. It can produce more inclusive, more caring, and less inward-looking forms of worker struggle against firms and regulatory bodies.

Though, as this chapter has shown, 'up-scaling' struggle beyond the national level can be a key way to defend and enhance wage workers' livelihoods, it would be wrong to think it an unalloyed good. There are two reasons why. First, even when workers (or consumers) in one set of places act with, or on behalf of, those located elsewhere, there are always some workers who are left out of the loop. No act of transnational solidarity can include *all* wage workers. There are simply too many for this to be a practical possibility in anything other than just rhetoric. Second, even just considering those who *are* involved in any given transnational worker initiative, there can be unintended and negative consequences of the actions taken. Let us take each scenario in turn by way of brief examples.

First, some acts of transnational labour solidarity are avowedly exclusive. They include one set of workers by defining their relevant constituency *against* a set of other workers. So the socio-geographic divisions within the global 'working class' do not just exist at the local scale but occur translocally too. The obvious examples here are the International Trade Secretariats, which represent different workers according to their industry. In itself this kind of industrial division is not necessarily bad, of course: the needs and rights of metal workers, for example, will not necessarily be quite the same as auto workers. Accordingly, it makes sense to have different organizations representing their respective interests. But in other circumstances transnational labour activism can be positively exclusionary, even competitive. A good example is offered by Herod (2001: 140–5). He recounts the post-World War Two efforts of American trade unionists to act on a transnational scale in defence of their domestic interests. Specifically, he looks at the activities of the AFL-CIO in Latin America from the 1950s onwards. Among other things, these activities involved building alliances with trade unions from Mexico to Argentina. But, as Herod explains, these alliances were very selective. Operating in a climate of Cold War fear and American economic growth (at least until 1989 and 1974 respectively), the AFL-CIO tended only to support 'moderate' Latin American unions that were not 'tainted' by the ideas of communism. Indeed, more than this, Herod argues that the union actively tried to weaken unions that seemed 'anti-American' and unreceptive to the import of US goods. In this sense, the AFL-CIO was frequently an *ally* of US businesses, laying the ground for American investment in, and exports to, Latin America. The kind of transnational solidarity it built between the United States and non-US workers was thus a very partial one, predicated on excluding those Latin American workers considered too radical for the AFL-CIO's taste.

If Herod's discussion shows how borderless bridge-building can be exclusionary, other cases show that some workers lose even when they are *in* the loop, as it were. As Johns and Vural (*op. cit.* 1208) explain, in relation to the SSC, 'plant closings and job loss for impoverished workers in the underdeveloped countries may be an unintended result ...'. For example, when the Walt Disney Company was criticized for the dismal labour conditions in 14 of its Haitian subcontractors, H.H. Cutler – the largest subcontractor – announced it was pulling out of Haiti altogether because of a 'slump in demand' caused by criticism by the SSC and others. This would lead to 2,300 lost jobs in Haiti. Of course, the SSC's intention was not to close down garment suppliers in developing world countries where jobs and investment are much needed. Rather, the SSC argued that Disney should set a standard for other companies to follow, by insisting that Cutler and other contractors raise wages and improve conditions. The point, of course, is that the campaign could not *guarantee* workers in Haiti and elsewhere that this standard would be set. Firms like Disney and Cutler were and are free to ignore transnational worker activism and seek out places where such activism is absent. So supranational labour organizing comes with risks for those it aims to help.

SUMMARY

This chapter has explored the several ways in which workers and their representative institutions can up-scale their actions in order to defend or enhance their livelihoods. These scale-making initiatives are increasingly transnational in scope as a response to the enlarged terrain over which many firms and regulatory organizations now operate. We have seen examples of successful transnational organizing by workers led by union and/or non-union bodies. But we have also seen that an uncritical celebration of labour transnationalism is naïve. It is likely – indeed desirable – that initiatives like the Ravenswood campaign and the SSC will proliferate in the years to come. But we should end this chapter by asking a key question: under what conditions will such initiatives, realistically, achieve their self-stated goals? To answer this question we might reflect on the differences between national unions and transnational worker organizations. The longshoremen example considered in this chapter is just one of literally thousands of cases where a strong national union has acted successfully with and for its membership. In these cases the influence of national unions derives in part from their ability to reach into the centres of business and regulatory power, themselves organized nationally. In the UK, for example, for many years national unions were actively included in CBI (Confederation of British Industry) and government deliberations on economic matters. By contrast, at the current time the supranational institutions of labour are out of step with those of business and regulators in one crucial respect. Though these institutions can now match the *scale* of the latter two, they do not necessarily have the same *power*. For example, unions like the ICFTU and the WFTU simply do not have the clout of a body like the WTO, nor do they have an 'open door' to its policy deliberations. This is unfortunate because the WTO has a major impact on workers worldwide when setting trade rules. The reason for this relative lack of power is that the ICFTU and the WFTU cannot readily mobilize workers on the ground to act against employers and regulators. It takes time for decisions taken by a global union body to filter down to national and local union organizations. What's more, a slow process of persuasion might be needed to get workers in different countries to act in concert. European Works Councils (EWCs) are perhaps the current exception that proves the rule. These international worker bodies have rapidly (i) connected workers across Europe and (ii) had an influence on firms and regulators in Europe (see Wills, 2001). The reason, arguably, is that EWCs were formally set up under the auspices of the European Union and thus were, from the start, linked to the central power structures of European politics and business. No such equivalent body has been established by the WTO. Though the International Labour Office is a body of the United Nations, and thus a part of one of the world's best-known quasi-state organizations, it does not have a strong voice. Its remit does not give it especial power either within the UN or in other global business/regulatory bodies like the WTO or the IMF. Thus, despite its many admirable pro-worker charters and declarations (see Hensman, 2001), the ILO is a dog that barks but cannot bite. The lesson is this: if the victories won in Ravenswood and elsewhere are not to be isolated examples, then the labour movement needs to create transnational institutions that have a *formally recognized seat* at the global bargaining table with firms and regulators. Whether this eventuates only time will tell.

Part Four

WORKERS, GEOGRAPHICAL SCALE AND IN/JUSTICE

nine

The Geographical Dilemmas of Justice

In the previous eight chapters we have tried to show how and why geography matters to workers and workers to geography. After a scene-setting Chapter 1, Chapter 2 explored the social dimensions of labour, while the rest of Part One inserted these into a geographical landscape of difference and similarity, of uneven development and reciprocal dependence. The concepts of place, space and scale, we argued in Part One, help us understand just what the problems and possibilities for wage workers are in the early twenty-first century. In Parts Two and Three we explored these threats and opportunities by way of case studies drawn from around the world. But where does all this theoretical and factual information leave us? What broader lessons can we learn, as both interested citizens and actual (or potential) wage-workers ourselves, from the preceding chapters?

In this concluding chapter we answer these questions in three stages. First, we draw attention to the geographical dilemmas touched upon in the last chapter. We do so because it seems to us that the more interdependent workers become, the more the tensions between local agendas and translocal agendas will be amplified. These tensions present some real difficulties for efforts to improve the lot of wage-workers and other groups dependent on the vagaries of global capitalism. They must be confronted head-on if any progress is to be made in somehow balancing worker needs and wants at different geographical scales. Second, this leads us to explore what kinds of rallying points there might be for wage-workers worldwide to unify around today. As the 'working class' has grown numerically and fractured internally (along the lines of gender, 'race', place, etc.), it is imperative that some powerful new unifying foci be developed in order to take further the kind of *global* organizing that would truly empower workers to curb the worst excesses of capitalism. This is really a question of finding new commonalities behind the manifest differences among working people worldwide. Finally, we reflect upon our own positioning as authors of this book and ask readers to reflect on their positioning too. The issues discussed in this book ultimately affect *all* of us. We cannot refuse our connections with distant strangers. The challenge is not just to acknowledge them but to *understand* them. Recalling Marx's dictum, only then can we change the world for the better.

UNRESOLVABLE PROBLEMS?

In Alexander Dumas's famous story of the *Three Musketeers* the heroes of his tale have a rousing refrain which goes 'All for one, and one for all!' Rather like Dumas' musketeers, many contemporary workers and allied social groups cannot afford to think and act alone if they are to avoid suffering real work-related injustices. As we have seen throughout this book, the fact of place interdependency at a number of geographical scales gives workers little choice but to organize translocally. This kind of translocal organizing does not, of course, necessarily mean that workers in different places are struggling for the greater good of all of them. In Chapter 8, for example, we saw how workers sometimes harness translocal forces in the interest of resolutely local agendas. Thus the well-known saying, 'United we stand, divided we fall!' is not strictly true in the case of all contemporary wage workers. Some workers in some places can certainly benefit from acting alone. The defence of local interests can, as we have seen, often be achieved by internalizing the competitive ethic of capitalism and acting against workers in other parts of the world. Yet, as the previous chapter showed, there are some real benefits to be derived from workers (alone and with non-labour interests) acting in concert over space. This brings us to two crucial issues: the barriers to geographical cooperation among workers are in play in our capitalist world; and whether these barriers can be overcome or whether cases like those discussed in Chapters 7 and 8 are exceptions that prove the rule? In short, we need to ask what is at the root of the geographical dilemmas confronting workers worldwide and whether they can be successfully negotiated.

Universals and particulars

We can begin to get a handle on these issues by considering David Harvey's (2001: Chapter 9) essay 'Militant particularism and global ambition'. In it Harvey reflects upon his involvement in a local campaign to save jobs in a Rover car factory at Cowley near Oxford, UK. This involvement brought Harvey into contact with a labour activist and researcher called Teresa Hayter, who was committed to defending jobs in the plant. Recalling a lunch with Hayter during the Rover workers' campaign, Harvey was challenged by her to define his loyalities. Did they lie with the local workers and the local community who depended on Rover for jobs? Or was he 'a free-floating ... intellectual who had no particular loyalties to anyone?' (2001: 160). Harvey's answer to Hayter's question, he continues, were 'yes' and 'no' respectively – which is not to say he had *only* local worker/community interests at heart. As Harvey put it: 'I was concerned about the incredible overcapacity in the automobile industry in Britain, and in Europe ... [W]e had to find some way to protect workers' interests in general without falling into a reactionary [localism] ... But across what space should that generality be calculated?' (2001: 161). In other words, Harvey's response to Hayter was to argue that defining workers' interests at the local scale alone was *necessary but insufficient*. Saving jobs at Rover, while it would benefit the Cowley community, would do nothing to further the interest of autoworkers as a whole in a competitive climate where overproduction was threatening the livelihoods of workers in many car plants across Europe.

Harvey's attempt to balance local needs with transnational needs reveals the geographical dilemmas confronting contemporary workers worldwide. At what geographical scale and with which people do the interests of any given set of workers lie? Somehow local needs and wants must be made to complement, rather than contradict, national, international and ultimately global needs and wants. For the Marxist intellectual Raymond Williams, one of the traditional strengths of the labour movement has been to forge inter-place ties among diverse working groups of just this kind. As he put it (cited in Harvey, 2001: 172):

> The unique and extraordinary character of working-class self-organization has been ... to connect particular struggles to a general struggle in one quite special way. It has set out, as a movement, to make real what it first sight is the extraordinary claim that the defence and advancement of certain particular interests, *properly brought together*, are in fact the general interest.

However, despite Williams' optimistic appraisal, the problems that workers and allied social groups face in connecting local and translocal agendas are truly formidable. To begin with, because of the objective and subjective dimensions of place explored in Chapter 3, it is always easier for workers to define their loyalties locally rather than at broader geographical scales. It can be very difficult to feel ties of solidarity with distant workers one never sees; it's far easier to identify with those one works with day-in and day-out. Charity, as the saying goes, usually begins at home (and often ends there too). Second, the fact of uneven geographical development insinuates real inequity into the landscape of labour. Should workers in the developed world allow jobs and investment to go to places in the developing world in a spirit of benevolence? Can developing world workers selflessly refuse foreign direct investment because they want their developing world brethren to keep their jobs? Third, though it may seem paradoxical, most wage-workers do not, in fact, belong to the kinds of union and non-union bodies discussed in earlier chapter. These bodies are the principal institutions that build inter-place worker connections, as we have seen. But they scarcely exist in many parts of the Third and Second worlds, while in the developed world membership of such bodies has dropped in many countries (see Box 9.1). Finally, the world's wage-working population is irreducibly diverse in both the social and the geographical sense. Differences of culture, ethnicity, language, dress, nationality, religious beliefs and so on can act as real barriers to understanding and cooperation between spatially separated working communities. A tragic example of this is the attitude of extreme right-wing Germans to immigrant workers. Xenophobic neo-Nazis have for many years targeted *gasterbeiter* (guest-workers) from Southern Europe and North Africa for racial attacks in an effort to keep German cities ethnically 'pure'.

Box 9.1 Trade union membership worldwide

Trade union membership has varied significantly over time and space. After the Second World War membership of a trade union was the norm for many workers in the 'developed' world. As we saw in Chapter 8, national unions were the cornerstone of the labour movements in countries in Western Europe and North America, securing better terms and conditions of employment for workers. Equally, until the demise of communism in the USSR and Eastern Europe, union membership was also high in the many 'non-capitalist' countries that existed during the so-called Cold War. However, outside these geographical regions, governments and employers have often resisted the formation of unions. In the 'developing' world workers have not generally enjoyed the same level of protection, and have had to engage in long and often violent struggles to be able to establish unions. Yet as trade unions have come into being in some developing countries, elsewhere in the world union density – that is, union membership as a percentage of workers – is in decline. This is especially the case in G7 countries such as France, the US and the UK (where two decades of economic restructuring and right-wing governments have undermined union power) and in the former communist bloc (where the collapse of state-socialism in the late 1980s left an organizational vacuum among working people). Where once almost two-thirds of workers were members of a labour union in these countries, density rates have now dropped to just over 25 per cent in some cases. As we can see below, in some countries unions can no longer claim to speak for the majority of workers, posing a direct challenge to their legitimacy and challenging labour unions to re-think their strategies, for example in the ways we have seen in Chapters 6 and 8.

Country	*Level of union membership in*	*Level of union membership in*
Australia	1980 – 55%	1998 – 28.1%
Canada	1980 – 37.6%	1999 – 30.1%
Germany	1991 – 34.8%	1998 – 26.1%
India	1980 – 16.7%	1997 – 26.2%
Ireland	1988 – 48.4%	1996 – 43.6%
Japan	1980 – 30.8%	2000 – 21.5%
Rep of Korea	1980 – 21%	2000 – 12%
New Zealand	1985 – 43.5%	1998 – 17.7%
Singapore	1980 – 26.8%	2000 – 18%
Slovakia	1996 – 77.9%	1999 – 59.8%
UK	1982 – 49.9%	1998 – 33%
US	1983 – 20.1%	2001 – 13.5%

Source: ILO Bureau of Statistics (2003)

Several of the above mentioned reasons that militate against interplace cooperation among workers are more worrying than others. These particular barriers to geographical solidarity relate to what are called *incommensurables*. Incommensurable are things or ideas that cannot be compared on a common scale of reference. That is, the differences between them cannot be reconciled. Thus if one person believes that God made the earth in seven days and another that it evolved from the Big Bang then they hold incommensurable views on natural history. Likewise, it is important to recognize that even with the best will in the world, workers may not be able to reconcile their myriad socio-geographical differences to pursue a common cause. As one commentator puts it (Holloway, 2002: 29), 'good people can disagree about the nature of the good itself'. In other words, in certain situations it may not be possible to deliver on William's – and Harvey's – aspirations for a translocal worker solidarity. The 'extraordinary claim' that the defence and advancement of certain particular interests are in fact the general interest may be *so* extraordinary as to be infeasible. This is especially the case now that the aforementioned diversity of wage workers worldwide is unprecedented. As Hyman (1999) argues, the seemingly limitless power of capitalism to bring peoples of all ages and beliefs within one economic universe is outstripping the capacity of unions and other pro-worker organizations to identify genuine commonalities within these differences. For example, what 'common' interests, realistically, do a Burmese sweatshop worker and a Melbourne stockbroker possess? Can their 'local worlds' somehow be made commensurable in a more general project to improve the livelihood of the former without undermining the right of both to lead different lives? Can a female IBM executive in Los Angles feel solidarity with her Filipina immigrant child-minder? Regrettably, the answer is often 'no'. Many wage workers, for understandable reasons, remain immured in their particular socio-geographic envelope and are unwilling or unable to 'negotiate through difference and similarity to formulate collective strategies without sacrificing local loyalties and militant particularisms' (Swyngedouw, 1997: 176).

What all this emphasizes is that attempts by workers to develop transnational solidarities require a major ideological and organizational *effort* on their part. A wider inter-place worker consciousness does not automatically spring forth simply because wage workers share a common class position within the structure of global capitalism. Rather, it has to be actively constructed in both thought and practice. One the one side, workers need an idea, an issue or an identity that they can rally-around *en masse* (the subject of the chapter's next main section). On the other side, as Chapter 8 showed, workers need the organizational means to make such an idea, issue or identity flesh. Without such organizational capacities, noble thoughts about transnational worker cooperation will remain just that: thoughts that have no real world impact. Even when this combination of issues/ideas/identities and organizational means is achieved, it is important to recognize that reconciling local worker agendas with transnational ones will rarely be a smooth process. Compromises must usually be made. This is what Jonas (1998: 325) calls the 'local-global paradox'. This paradox arises because what is 'good' for one set of workers at one geographical scale may not be good for other workers at other geographical scales. The balance of gain and loss is very much case specific, but is the essence of translocal worker struggles to confront this difficult balancing act, willingly or not.

Reasons to be cheerful: connecting difference

All is not doom and gloom, however. Many worker differences *are* commensurable: they can be made to speak the same language, as it were. As a prelude to discussing what kinds of big ideas, common issues or constructed identities the labour movement might rally-around now and in the future, it is worth pinpointing the organizational tools currently at the movement's disposal. In the first place, there exist today an unprecedented number of macro-regional, international and global worker organizations, from the ILO to the ICFTU to the Southern Initiative on Globalization and Trade Union Rights. In time, these might usefully consolidate and combine into a few powerful 'super' unions. Second, these organizations have helped to put labour transnationalism 'on the agenda' in many national trade union bodies and in many non-labour organizations in civil society. They force these unions to 'look up' beyond the national scale where they are not already doing so. Third, because transnational organizations like the ICFTU have existed for several decades, there is a history of international, inter-place organizing for contemporary workers to draw upon and learn from. Fourth, the very same telecommunications and transport networks that have allowed many firms to be geographically footloose have also strengthened workers' capacities to act transnationally and to act quickly where necessary. Ours is a world of 'time-space compression' (Harvey, 1989b). The 'shrinking' of the world caused by email, the internet, the global media, supersonic aircraft and so on have bolstered the possibilities for truly global forms of worker cooperation. Information can be exchanged in the twinkle of an eye and actions in many different places can be readily coordinated in real time. Fifth, as some of the examples in the previous chapter showed, the fact that many firms are today highly 'networked' across space means that they are very vulnerable to actions that interrupt the flows of goods and information upon which they depend. Finally, like it or loath it, the fact that one language – English – is today become the *lingua franca* worldwide arguably helps the cause of labour transnationalism. It will increasingly figure as the oral currency that different local and national worker bodies can deal in when confronting one another over issues of common concern.

CONTINGENT UNIVERSALS: FINDING A COMMON GROUND

We now arrive at a fundamental question for contemporary workers, their representative institutions and allied groups in civil society: what issues, ideas or identities can otherwise different workers worldwide rally around for the greater good of all? We phrase the question this way because these rallying points ultimately need to be used not to *eliminate* all differences between wage workers but to do away with 'negative differences' (like poverty), while protecting 'positive differences' (like the right of different places to make different commodities in different ways). What analysts like Harvey and Williams are arguing is not that all forms of socio-geographic diversity be overcome in some homogenous worker struggle but that the best elements of diversity be defended *through* a commitment to broader goals. To repeat Hyman (1999: 99) from Chapter 4: 'without differentiation there would be no need for solidarity'. In light of this, we need to consider different types of worker difference before considering how they can be erased or nurtured through a broader global struggle.

Promoting difference through similarity

In their book *A World of Difference*, American geographers Eric Sheppard and Philip Porter (1998: 30) argue that socio-geographic difference 'is at once a blessing and a problem'. It is a blessing, they argue, because it is depressing to imagine a world in which all peoples and all places are more or less the same. Differences – of culture, religion, food and so on – enrich our lives, and when they are fused in novel ways can lead to unexpected and equally valuable forms of hybrid difference. However, as Sheppard and Porter observe, not all differences are equal, as it were. For instance, the kind of 'differences' that make certain workers in certain places vulnerable to unemployment, low pay, harsh working conditions and so on are clearly not ones most people would wish to preserve. But, to complicate matters, we need to understand that virtually all socio-geographic differences are *relational*. That is, these differences are actively produced through relations of interconnection and interdependence – precisely the kinds of relations we have emphasized throughout this book. This has two implications. First, it means that differences among wage workers worldwide do not bootstrap themselves into existence *sui generis*. There are no 'pure' differences between working groups in a global capitalist economy. Second, it follows that wage workers and communities in one set of places are partly *responsible* for producing differences among other workers in other places. Inter-place competition for jobs and investment is the most obvious example of this and one that we illustrated in Part Two. Such competition produces 'winners' and 'losers', which is another way of saying uneven geographical development. Because the winners stand to lose if they forgo such competition in place of a cooperative, non-competitive ethic, then it is plain to see how difficult it can be to eradicate 'negative' inter-place differences between workers.

The challenge for the labour movement and allied organizations is thus a demanding one. On the one side, cooperative inter-place ties need to be forged at several scales that can eliminate the worst forms of socio-geographic difference. On the other side, 'positive differences' must somehow be preserved in this process, otherwise a bland socio-geographic levelling ('up' or 'down') might result. The line separating 'positive' and 'negative' differences is a fine one. One of the reasons we ended Chapters 6 and 8 with a discussion of what constitutes 'progressive' and 'regressive' forms of worker struggle was to ram home this important point. Likewise, our concluding discussion of migration in Chapter 7 was designed to emphasize the perspectival nature of evaluation. What seems good to one set of workers in one place, can look quite the opposite to others living elsewhere or in the same locality. How, then, to square the circle? One answer is to say – quite rightly – that while some differences can be evaluated in a *relative* way (depending on the perspective of the evaluator) others demand a more *absolute*, uncompromising stance. For instance, few would find the long-term unemployment of groups of wage workers – with all the consequences for place, personal well-being and community life this entails – acceptable. Likewise, when certain workers are compelled to labour in neo-Dickensian sweatshop conditions, this is the kind of difference many people elsewhere – as the case study of consumption politics showed – are not prepared to tolerate. In short, wage workers can and should come together to lay a common foundation on the basis of which positive differences can flourish. But what will this foundation look like?

New rallying points

In recent years the labour movement, alone and in alliance with non-labour interests, has gone some way to manufacturing new rallying-points for worker action worldwide. We use the word 'manufacturing' deliberately because, as we have already argued, lines of inter-place solidarity must be actively *made* – they do not automatically make themselves apparent. They are thus contingent universals: that is, commonalities that appear to be 'inherent' to all wage workers but which are, in fact, strategically useful concoctions. This need to fabricate common bonds is especially acute now that wage workers must, in many cases, cooperate globally if they are to further local and non-local interests. Given the worldwide scale at which many firms and regulatory institutions are organizing, macro-regional and even international scales of labour struggle may benefit many workers but leave still others vulnerable to the ills of capitalism. The aim, then, is to create *universal* rallying-points around which workers with widely different identities, occupations and experiences can *genuinely* and *willingly* come together.

As argued above, in theory these rallying-points can relate to any of three things: issues, ideas or identities. It is worth noting that the last of these is the most difficult to create, yet the one that would best guarantee global worker solidarity as and when necessary. Because identities are lived and visceral, they can motivate action at the deepest psychological levels. The difficulty arises because there are no convincing arguments that a 'core' identity really does cross-cut the myriad personal and group differences between workers worldwide. As we showed in Part One, worker identities are 'over-determined'. They are the complex and variable outcome of the multiple 'subject-positions' that different workers occupy in everyday life. In previous attempts to encourage transnational worker identities, class was seen as the unifying glue. Following Marx and Engels's famous injunction in the *Communist Manifesto* – 'Workers of the world unite! You have nothing but your chains to lose' – transnational worker organizations, until relatively recently, assumed that workers' common class position within capitalism could transcend socio-geographic differences. However, today it is abundantly clear that a 'working class' identity ignores and excludes all those important non-class elements of worker identity that make working people what they are. So class is a limited and exclusive basis for any form of global struggle founded on a common worker identity. Indeed, it is fair to say that virtually *all* attempts to locate such an identity risk falsely 'essentializing' workers, which is to say they risk mistaking a part of their identity as *the* fundamental element of that identity. This said, there are still some benefits to be gained from using identity as a global rallying-point. A good example is the attempt by Women Working Worldwide (WWW) – a UK non-governmental organization – to counteract gender discrimination against women workers. This organization strategically accents the fact that millions of workers share the identity 'woman'. Though, clearly, the WWW's agenda excludes male workers, it does so deliberately because it is precisely the 'female' element of women worker's identities that underpins the discrimination they suffer (see Hale and Shaw, 2001). Likewise, it is possible to found worker campaigns on the basis of *several* discriminated-against identities (in a 'rainbow coalition'), like those relating to certain genders, ethnicities and sexualities.

So much for a global solidarity founded on identity. What about one founded on ideas and/or issues? These are far more tractable bases for constructing worldwide solidarity. In practice ideas and issues are intertwined, but we can treat them separately for analytical purposes. Ideas matter because they can encapsulate the kind of people we are or want to be. Think of famous ideas like 'All people are born equal' or 'Treat others as you would want yourself to be treated' and the effects they have had. Unlike identities, ideas can be readily created. The task is then to have them accepted as worthy ideas that can, in the present context, inspire inter-place worker actions. Apropos the distinction made above between relative and absolute evaluations of difference, some of the current attempts to create compelling ideas for a global worker solidarity focus on the second type of evaluation. A good example is the ongoing effort by the ILO to use the famous Declaration of Human Rights created by its parent organization – the United Nations – to defend workers against the worst excesses of capitalism worldwide. The declaration, first made in 1948, argues that *all* humans, regardlesss of 'race', place, creed, gender, age, etc. have a set of basic and fundamental rights that must be respected. Though the declaration is not about wage workers *per se*, it can be used to defend their interests when and where their human rights are breached in an economic context. In other words, the ILO has tried to make labour issues human rights issues as a way of addressing abuses of workers and stressing the common rights all workers should properly be entitled to. It has done so through its 1998 Declaration on Fundamental Principles and Rights at Work. This move immediately raises the stakes against firms and regulatory bodies by making worker issues *more* than simply worker issues. The declaration is just one example of how a key idea can serve as what Marx called a 'material force' in the cause of paid labourers and their allies. The ILO also has over 180 declarations of its own designed to enhance the working lives of labourers worldwide (see www.ilo.org).

Finally, certain issues can, like certain ideas, serve as powerful unifying forces for global worker solidarity. By issues we mean particular problems that are commonly experienced by many workers or particular things that are widely sought after. Good examples are the minimum wage and the living wage. Quite simply, individuals, families and communities who depend upon wage labour for their well-being cannot lead a dignified life if their income is below what is required for physical reproduction. A certain base-line income is an absolute *sine qua non* for all wage labourers and their dependents. Because of the serious consequences of *not* receiving a minimum wage or living wage, this is an issue where it is possible to persuade many workers that 'An injury to one is an injury to all' – at least in theory. This is not to imply that transnational worker organizations should agitate for a *single*, uniform minimum or living wage worldwide. This would ignore the real differences in the cost of living between places and entail a levelling-up of wages that millions of firms and regulatory institutions would vehemently oppose (the so-called 'high road' goal of transnational labourism). But low-income workers and their better-off colleagues could plausibly unite in a project to ensure that locally and nationally adjusted basic pay levels are enforced (Galbraith, 1995).

Conceptions of justice: alternative moral geographies

Whichever rallying-points might be favoured, both now and in the future, it is clear that immense problems confront global worker organizing, notwithstanding the 'reasons to be cheerful' stated earlier. As we have explained in the previous pages, these problems are primarily scale problems (At what scale do and should wage workers define their loyalities?). In addition, we might also mention that many logistical problems remain (the practical difficulties of coordinating worker actions across time and space): the mere existence of advanced telecommunications does not, in itself, guarantee successful transnational worker campaigns. Rather than dwell on logistical issues though, we want to end this section by highlighting a third challenge: namely, deciding what conceptions of *justice* are most appropriately deployed in endeavours to forge a new global worker solidarity. It is a challenge that transnational pro-worker organizations arguably need to confront more squarely than they currently do. If, logistics permitting, new rallying-points can bring about global labour solidarity, the question still remains as to what *kind* of justice these rallying-points can deliver.

Why talk about justice? Indeed, what *is* justice? We began to sketch an answer at the end of Chapter 4. As Mitchell (2000: 288) observes, 'justice is an ideal against which to measure the accomplishment, the practice and the aims of any society'. When that ideal is offended against an 'injustice' can be said to occur. One of the virtues of phrasing ideals in terms of in/justice is that it is a powerful and emotive way of ensuring they are attained – or ensuring that someone speaks out when they are *not* attained. Thus the ideal of 'human rights', discussed above, lends itself readily to the discourse of in/justice. Indeed, for generations, wage-workers have used this discourse and attached it to any number of ideas and issues in order to make their voice heard. For instance, if we think of ideas, then aside from the notion of rights workers have also used elemental ideas of 'needs' and 'entitlements' as benchmarks for achieving justice within and beyond the workplace.

In the twenty-first century it is important that the labour movement continue using the language of justice and injustice to inform its struggles. It is a compelling, evaluative language that can inspire people to identify and resist unfair practices. The difficulty, though, is that justice is not a single, homogenous thing that all wage-workers can pin their colours to. It is, rather, a *plural* phenomenon: there are different forms of justice and thus different modalities of injustice. This would not be a problem for the labour movement – indeed, it could be an advantage – except that these types of justice are not always consistent with one another. So contemporary efforts to build genuine global solidarity among wage-workers must somehow wrestle with the question: what *kinds* of justice should it struggle for and will they be appropriate for all? The answer to this question has a geographical and a non-geographical dimension.

We have already discussed the former at length, albeit not under the rubric of in/justice. To recap, what is just at one geographical scale for one set of workers might be unjust for other workers elsewhere. For instance, if justice is defined as a situation where an individual can enjoy liberty then a well-paid group of Boeing aeronautical engineers in Seattle might feel they are justly treated by their employer. However, going back to our earlier discussion of responsibility, these engineers' liberty in this part of the Boeing industrial empire cannot be divorced from what is going on elsewhere in the company. Do, say, workers in Mexico stitching seat-covers for Boeing aircraft enjoy factory conditions and an income that affords them the same degree of liberty within or beyond the workplace?

So much for the geographical dilemmas of justice. These are further complicated by putatively non-geographical dilemmas. To illustrate the potential intractability of these dilemmas we can draw upon the important book *Justice Interruptus*, written by political philosopher Nancy Fraser (1997). Fraser argues that in the current era two important notions of justice are competing for the attention of those on the Left: namely *economic* and *cultural* justice. Both are directly relevant to the contemporary labour movement. Economic justice is a situation where all people receive a certain basic income that allows them to live a dignified life. Cultural justice is a situation where all people are allowed to express themselves without fear of discrimination. Examples of economic injustice would be long-term unemployment and the payment of poverty-wages to certain workers. Examples of cultural injustice would be paying women or homosexuals less than men or heterosexuals for the same job. Both types of in/justice, Fraser argues, intertwine and can be mutually reinforcing. But, she continues, they can also contradict one another. For instance, if we take the case of a person of colour who suffers both types of injustice simultaneously in a predominantly white workplace, we can conceive of two remedies for the situation. These are better pay (so that s/he is paid the same as white workers for the same work) and the fostering of a multi-cultural ethic by the person's employer. So far so good. However, as Fraser argues, this is not as positive as it seems. In this case, the pursuit of cultural justice – allowing the person of colour to celebrate their non-whiteness – can, ironically, contradict the aims of economic justice that would materially support that act of personal expression. Why? Because cultural justice rests upon ideas of social sameness – that everybody, regardless, is entitled to a certain income – while economic justice rests upon ideas of social difference – that everybody is not, by any means, similar. Fraser thus worries that the project of engendering cultural justice can inadvertently reinforce the kinds of competitive, non-cooperative ethic between people – in the case of this book, between workers and places – that an economic justice project wants to dispel. Even using such a simple example, it is clear that contemporary attempts to achieve justice for wage workers and their allies worldwide will have to grapple with the complementarities and contradictions between different notions of justice. There will be no easy solutions.

NORMATIVE STANDPOINTS AND CRITICAL RESEARCHERS

We started this book with a declaration and an invitation. Upfront about our Left-wing sensibilities, we invited readers to understand the geographies of labour, not for their own sake but as citizens who will one day become active players in the interdependent worlds of work we have explored in this book. So *Spaces of Work* has been written as an avowedly *critical* intervention. Like Mitchell (2000: xiv), we take this to mean two things. First, we are critical in the sense that we have taken a political position on the things we have analysed in this book. Second, though, we want readers to take a stance too, and not necessarily the one we have adopted as authors. This book is not a 'manual' designed to impart 'the truth' about wage workers. Readers should be critical of what we, as writers who view the world through particular lenses, have written. By way of a conclusion to this chapter – and to the book as a whole – we thus want to end by addressing two criticisms that may, quite reasonably, be levelled at the arguments we have presented in Parts One, Two and Three.

Status quo criticism?

The first criticism is that we are insufficiently critical! That is, some readers might charge us with 'status quo criticism'. To understand what this is we can distinguish between 'reformist' and 'revolutionary' forms of analysis and action. Karl Marx, one of the theorists who has inspired our analysis, was very much a revolutionary. That is, he wanted to *overthrow* capitalism, seeing this as the only way that working people and communities could enjoy a full and rewarding life. Clearly, we have not recommended the kind of transformative politics that Marx favoured over a century ago. Instead, we have tried to identify the kinds of *non-revolutionary* actions that wage workers can undertake at different geographical scales in order to improve the life conditions of some or many of them. We have departed from several other Left-wing labour analysts by refusing to make sweeping generalizations about the state of most wage workers today. Our argument has been that there is ample 'wiggle room' for workers in a capitalist word (albeit unevenly distributed) and that socio-geographical context matters. Workers are not 'dupes' but nor can they readily topple the capitalist mode of production; equally, understanding how and what workers do demands an attention to the contingencies of their social and geographical location. Does this mean that revolutionary action is impossible and/or undesirable? And does the kind of reformist action we have explored through case studies amount to an ineffectual 'tinkering' with the machinations of global capitalism?

To take each question in turn, it seems to us that revolutionary action is, at this moment in history, not a viable option for the labour movement. We said as much at the start of this book (see Preface). Notwithstanding the wider ground swell of opposition to capitalism in its neoliberal form – as seen in Seattle, Genoa, London and elsewhere – the forces ranged in defence of the system are just too strong to be dislodged. In part, as we have shown, these forces lie *within* the labouring classes themselves! Many wage-workers and their dependents enjoy real benefits living and labouring in a capitalist universe. These benefits cannot be underestimated. But even if revolution were an option – which is often is during moments of rationality/legitimation/fiscal crisis (see Chapter 2) – it would not deliver some post-capitalist utopia overnight (or, perhaps, ever). The twentieth-century experiments in communism (in the USSR and elsewhere) and recent thought experiments in imagining a socialist economy (for example, Sayer, 1995) both suggest that immense practical problems lie in the path of fashioning a non-capitalist future. In light of this, it seems to us that it is wrong to dismiss pro-worker action within the parameters set by capitalism as mere 'tinkering'. The case studies of worker agency we have presented each, in different ways, show the palpable and sometimes life-changing gains that can be made when labour fights back within the current economic system. A particularly powerful example of this, which we have not considered in this book, is the Zapatista movement in Mexico. Though not strictly a labour movement – it began life in 1994 as an armed uprising by indigenous (or first nation) peasants in the province of Chiapas, in southern Mexico – it shows what can be done when those marginalized by capitalist society hit back aggressively at those in power. In early 2001, after several unsuccessful military attempts to suppress the Zaptistas, the Mexican President formally acknowledged their right to have their concerns addressed by the national and provincial state. In this case, armed guerilla action by a discontented group was both necessary and successful in effecting positive action, albeit within the parameters of capitalism in Mexico (see Castells, 1997: Chapter 2).

Moral relativism and its alternatives

A second charge that some readers might level at us is that we are closet moral relativists. A moral relativist is someone who argues that there are no firm foundations for making judgements about the 'good' or the 'bad', the 'positive' or the 'negative'. Relativists argue that moral judgements do not come from the skies in tablets of stone. Rather, they are made by individuals and groups who judge the world according to their personal and cultural mindsets or 'lifeworlds'.

Because these lifeworlds are plural, relativists argue that moral judgements must be plural too (Walzer, 1983). The embrace of moral relativism can be liberating. It allows us to acknowledge the dangers of ever supposing that there is one 'correct' standard for evaluating the world's multifarious joys and dangers. But it can also, in its extreme forms, lead to the abyss of 'anything goes'. Thus, because many Nazis thought that killing Jews was morally 'acceptable', an extreme relativist would be forced to respect this viewpoint.

We, as authors, would certainly not want to endorse extreme moral relativism. But we have, clearly, argued for a weaker form of relativism in the preceding pages. At various points we have argued against those Left-wing analysts who try to make absolute judgements about the pros and cons of worker action at certain geographical scales. One of our aims has been to sensitize readers to the variety of perspectives within the global 'working class' depending on who and where the workers in question are. This variety is irreducible. It cannot be wished away or denied. In drawing it to readers' attention we have engaged in what moral philosophers call 'descriptive ethics'. That is, we have tried to describe the various actually existing moral judgements existing out there in the world. But this still begs the question of *our own* judgement, as authors, of the judgements that differently positioned workers make.

This is a question of what is called 'normative ethics'. Do we accept the variety of worker judgements of the 'good' and the 'bad' of their actions and circumstances? Or do we, as analysts, criticize some or all of these judgements (and if so on what grounds)? The sociologist and geographer Andrew Sayer (2000) argues that it is irresponsible for critical academics to remain agnostic about the things they analyse. We agree. Though we do not have the power to change the world directly (like, say, governments do), academics nonetheless have a valuable role to play because they have the training to offer in-depth investigations of what is going on outside the precincts of the university. The same is true of many journalists and professional writers, like George Monbiot, Jeremy Seabrook, Naomi Klein, Noreena Hertz, Michael Moore and John Pilger (see Box 9.2). So what is our critical standpoint? What, to put it differently, is our normative argument (that is, our argument about what *should* be done by wage workers and their allies as opposed to what they already do)? Our answer has coarsed throughout this chapter, but it as well to now make it explicit. First, it seems to us that some worker judgements of the 'good' and the 'bad' of their situation should be accepted as legitimate, but others strongly challenged. Specifically, worker actions at one scale that directly or indirectly cause injustices to other workers elsewhere should be the focus of criticism. Since we have acknowledged that justice is a relative matter, we reiterate that it is *fundamental* and *universal* injustices we are talking about here – like the breach of human rights. In a world where many locally-based workers are simply unaware that their local actions might have dire consequences for distant others, it is imperative to instil them with a geographical imagination. This is precisely what transnational unions and related worker organizations are now doing. This leads us to a second normative argument. It is that the labour movement needs to work very hard to allow 'positive' socio-geographic differences to flourish. What a global commitment to identifying universal ideas, issues or identities can do is to provide a secure basis for this to occur.

Box 9.2 Protest writers: the critics of global capitalism

Academics are one of three main groups who are able to publish close, critical analyses of global capitalism and the fate of workers within this economic system. The other two are journalists/independent writers and non-university researchers (for example, like those working for think-tanks and independent foundations). Academics are typically able to reach one very large but very particular audience with their writing: namely, students (many of the people reading this book). The other two groups are able to reach a much wider, more diverse public, mainly through books, and articles in widely circulating newspapers/magazines and reports. Famous independent writers who are highly critical of global capitalism include Naomi Klein, the columnist for the British *Guardian* newspaper George Monbiot, the freelance journalists John Pilger and Jeremy Seabrook, and the iconoclastic writer and film-maker Michael Moore. A noted writer who works for a non-university research institute is the American polymath Jeremy Rifkin. His Washington-based Foundation on Economic Trends allows him to investigate all manner of injustices in the world, some of them relating to wage-work (see Rifkin's 1995 book *The End of Work*). All of these critics of global capitalism – including the authors of this book – might be classed as 'protest writers'. That is, they call into question what seems 'normal' and 'acceptable' and encourage people to change the world for the better.

There's much to be done and much that it is already being done. There are many inspiring stories we have not had the space to recount here. We close, though, with a challenge to our student readers. What are you going to do when you enter – if you have not already entered – the world of paid labour? How will *you* resolve the tensions between your personal workplace ambitions and those of your working brethren? At what geographical scale will *you* define your loyalties when called upon to do so? We hope this book has provided you with the tools to answer these and other difficult questions in a considered and reasoned way. At the very least, *Spaces of Work* should encourage you to think – and maybe act – translocally when considering your own local working existence. As the great Marxist historian Eric Hobsbawn (2002: 12) puts it in his book *Interesting Times*, 'The world will not get better on its own.' In making it a better place, we should, as Neil Smith (1991: 178) insists, 'be planning something geographical'.

Glossary

This glossary offers short definitions of many of the key terms used in this book. Though some terms are defined in the text, we repeat them here for the sake of completeness. The definitions are specific to this book and are not necessarily those put forward in the wider literature listed in the bibliography. For fuller definitions readers are referred to both *The Dictionary of Human Geography*, 4th edition (Johnston et al., 2000), and *The Blackwell Dictionary of Social Thought* (Outhwaite et al., 1994). Full details of any references cited in the glossary can be found in the book's bibliography.

Capitalism
A specific and increasingly dominant mode of producing goods and services. Within capitalist societies commodities are produced for profit under conditions where competition among rival firms induces rapid technical innovation and unemployment. To describe something as 'capitalist' is to say that it's partly or wholly caught up in processes of economic competition, innovation and profit-seeking.

As political economists like Marx, Joseph Schumpeter, John Maynard-Keynes and John Kenneth Galbraith have shown, capitalism is a crisis prone economic system. Change is a constant in capitalism and disequilibrium the norm. Why is this? The three things comprising the 'logic' of capitalism – namely, growth, competition and innovation – are contradictory. The necessity for growth and the forces of competition together compel leap-frogging innovations within and among industries. New products appear daily – mobile phones, solar powered houses, wide-screen TVs, car valet services, or genetic tests – and existing products are constantly improved (witness how established automotive models are constantly restyled by their manufacturers). However, there are two problems with this process. First, competition among firms leads not only to innovation and growth but to economic 'wastage'. Some firms fall by the wayside as their economic rivals make better and/or cheaper commodities. These firms may close down or be swallowed up by others. Second, competition also leads, in many industries, to a tendency to reduce or hold steady the numbers and salaries of workers. Witness, for example, how 'smart machines' have come to do so many tasks previously undertaken by paid workers. Consequently, there is a structural tendency within capitalism to generate unemployment and to hold down the incomes of consumers (since workers are also buyers of commodities).

Because of these two internal contradictions, capitalist societies suffer periodic 'crises of over-accumulation'. Here production facilities and labourers lie idle in a condition where there is a glut of commodities. These crises can be local/regional, national, international or global. It very much depends on the number and type of firms within the capitalist system that are affected at any given moment (see Harvey, 1985: 128–35).

Capitalists
A generic name for those people who own the means of production and who employ workers to undertake certain jobs in return for a wage.

Class
A very complex term that normally refers to the differences between people in terms of their economic circumstances. Class has both objective and subjective dimensions. In the first sense it refers to people's location within the class structure of society. In capitalist societies there are two major economic classes – employers and employees – but in turn both are divided into **class fractions**. In the second sense, class refers to people's own awareness and definition of their class. Often, an individual may consider themselves to be a person of a certain class, when this is quite at odds with their objective class position. In recent years, some analysts have talked of an 'under-class' in capitalist societies, consisting of the long-term unemployed, the under-employed and the very low paid. In both its objective and subjective dimensions class is never isolated from non-class axes of social difference, like gender. Though we can identify class analytically, in practice it intertwines with other social relations.

Class consciousness
The awareness of members of a class that they belong to a class group. This consciousness may exist at the level of **class fractions** or of classes as a whole (**class-for-itself**).

Class-for-itself
The 'subjective' dimension of class whereby an individual becomes conscious of their **class position** and sees themselves as a member of a wider class group (**class-in-itself**).

Class fraction
A specific part of a wider **class-in-itself**. Class groups are internally divided by skill level, income, occupation and so on. These various axes of differentiation have a fuzzy order that produces an array of class fractions (for example, lower-middle class, upper-working class, etc.).

Class-in-itself
The 'objective' dimension of class that defines an individual's **class position** regardless of whether or not the individual understands or acknowledges it.

Class position
The specific class or class fraction an individual belongs to. Individuals can, through education, training, luck or family ties change their class position over time, but this is rarely easy or rapid.

Concessionary bargaining
A process where given sets of wage-workers are on the back-foot vis-à-vis their employers. These workers must thus make concessions – in wages, workplace rights, benefits and so on – if they are to retain their jobs.

Cross-class alliances
Temporary combinations of different classes or fractions of classes to pursue a mutually beneficial goal. Alliances are formal and explicit and usually arise in order to pursue specific goals. They can be forged at a variety of geographical scales.

Downsizing
A polite term for the process whereby firms radically rationalize their production activities by shedding jobs, reducing plant, selling off parts of the firm and so on.

Economic sector
A grouping of firms that produces the same general types of commodities (for example, the manufacturing sector). Economic sectors

are closely interrelated and in practice it is often difficult to distinguish them.

Employment relations
The social relationships that form between workers and employers surrounding the sale, purchase and use of **labour power**. These relations can take a variety of forms, from friendly and consensual to antagonistic, depending on the firms and workers in question.

Exchange value
The monetary value derived from the sale of a commodity. Commodities can be fixed (for example, a factory) or unfixed (for example, a car). According to some Marxist theorists, firms tend to value places less for their **use value** and more for their exchange value. This does not imply that places are literally for sale! Rather, it means that firms will be preoccupied with the profitability of locating production in one place versus another.

External labour markets
The sale and purchase of labour power where those *outside* a firm, workplace or economic sector can access – actually or potentially – certain jobs (see **internal labour market**).

Firm
An organization that produces commodities. Firms come in many shapes and sizes, from transnational corporations to local businesses. A firm consists of one or more **workplaces**.

Geographical scale
This is the historically variable scale at which specific sets of **social relationships** and institutions are arranged. Social relationships and institutions do not pre-exist the scale of their expression. Rather, their scale is *part* of how they work. Many social relationships and institutions work at and through a variety of scales, while others are highly scale specific. Geographical scale differs from cartographic scale. The latter provides a fixed metric within which things can be located and represented (as on a map). By contrast, geographical scales are socially and temporally variable. They are the geographical 'medium' that both enable and constrain how individuals, groups, social relations and institutions function. In this book we talk of 'sub-local', 'local', 'national', 'regional', 'international' and 'global' scales. However, these are not given in nature and the 'content' of these scales cannot be pre-determined.

Geographical divide-and-rule strategies
A process where certain firms or regulatory institutions play off workers in different places against one another for jobs and investment. This process is sometimes known as 'whipsawing' or the 'race to the bottom'.

Geographical imagination
A way of thinking that stresses how place and space – that is, geographical scales – are woven into the very fabric of social relationships and everyday life.

Geographical strategies
The explicit use of geography by firms, workers, regulatory institutions or actors in civil society in order to further their agendas. These strategies can, for instance, involve workers expanding the **geographical scale** of a strike in order to bring more workers into the struggle.

Globalization
A complex term that admits of no single definition. Generally, it has been used to refer to a increase in some or all of the following: the intensity, extensity, impact and velocity of social, economic, cultural, political and financial relationships between different places worldwide. In this book we forego using the term as an analytical category and focus instead on the way the concept has been used by and against workers.

Hegemony
A concept first developed by Italian Marxist Antonio Gramsci. Gramsci argued that social power is most effective when it does not take the form of physical force or coercion. Hegemony describes a process whereby dominant social groups pursue their agendas through subtly limiting the terms of debate with subordinate groups. Hegemony is about the forging of consensus about specific sets of 'acceptable' practices and behaviours by dominant and subordinate groups. The precise nature of hegemonic ideas, norms and habits varies historically and geographically.

Industry
A generic name for any act of commodity production or a grouping of firms undertaking the same types of commodity production (for example, the hotel industry).

Informal work
Work (usually paid) undertaken outside the normal rules, regulations and norms of the formal economy.

Internal labour markets
The sale and purchase of the labour of workers *within* a workplace, firm or economic sector. Internal labour markets involve workers *already* employed in a workplace, firm or economic sector rather than those coming in from the outside. Thus a shop floor worker in a factory may be promoted to a supervisor. These labour markets are thus partially 'sheltered' (see **external labour market**).

Just-in-time (JIT) production methods
During the 1980s US, Japanese and European firms introduced what are known as just-in-time production methods. These involve reducing component stockpiles to almost zero and reorganizing the production process so that parts arrive at the assembly factory just before they are needed – or 'just in time'.

Labour
Both the act of **work** and the name for people who do that work. In this book labour is mostly used in the second sense, referring specifically to *capitalist* workers. Until a few years ago, labour analysts tended to assume that workers were adult males (the so-called family 'breadwinner'). However, this classic image of workers is now badly outdated. Capitalist workers today are male and female, old and young. There are several varieties of labour that are capitalist and non-capitalist depending, as follows:

- The *self-employed* are just that: they work for themselves and their dependents:
- *Household* or *domestic work* is similarly controlled by the worker or their family/co-dependents and involves no monetary payment.
- *Volunteer work*, as the name suggests, is typically non-renumerated but, unlike domestic work, normally performed outside the home.
- *Communal work* involves a community organizing its collective labour to achieve certain agreed ends; it can be paid or non-paid.
- *Non-monetized work* involves an individual, individuals or groups exchanging non-monetary goods in payment for work done (so-called 'payment in kind').
- *Monetized work* involves an individual, individuals or groups exchanging money in payment for work done.

Labour demand
The requirement for a certain quality and quantity of workers among employers. Labour demand varies over time and across space and between firms and economic sectors.

Labour market
The buying and selling of **labour power**. This usually occurs at the local scale (see **local labour market**). However, if one abstracts from local labour markets and aggregates the number and type of labour market exchanges at any given moment one can talk about *the* labour market. This is not a market in the physical sense (as when one talks about buying fruit from the village market).

Labour market intermediaries
'Intermediaries' act as a bridge between employers and workers. Through their activities they attempt to match labour demand with labour supply, providing a range of services to employers and to workers. The US economist Paul Osterman (1999: 134) classifies intermediaries into three types: (i) 'one-on-one' intermediaries that passively accept job orders from firms and match these orders to those who have registered with the intermediary (such as the local job centre in the UK); (ii) intermediaries who are more active and aggressive in attracting employers ands workers (such as the traditional temp agency); and (iii) intermediaries who bargain with both parties, using their position in the labour market to change the way labour is sold within it (such as high-end job search companies). In his work in Silicon Valley, Benner (2002: Table 3.2) distinguished between three different types of labour market intermediaries based on the way they were organized:

Organization type	*Examples*
For-profit sector	Temporary staffing agencies Online job search sites
Membership based	Union-based initiatives Membership-based associations (such as coops)
Public sector	Job centres Voluntary organizations (such as those who organize work for ex-service men and women)

Labour movement
The various workers and worker organizations who, together, actively try to advance worker interests at a variety of geographical scales. The

labour movement is organized around trade unions primarily, but an increasing number of non-union bodies are now important too.

Labour power
The capacity to undertake **work**. In capitalist societies, workers sell their labour power to employers for certain periods of time.

Local labour market
Labour markets do not exist on the head of a pin. Rather, they are normally local in **geographical scale**. This means that the supply of workers and the performance of work takes place within a circumscribed area on a daily basis. This is partly because workers must return home at the end of each working day.

Local scale
The relatively small **geographical scale** at which daily life is acted out. Synonymous with **place**, the local scale is not fixed in area. What's 'local' will vary in size and content from situation to situation.

Locality dependence
Those social relations, institutions and practices that necessarily have an exclusively or partly **place-based** dimension to them.

Local growth coalitions
Alliances between **place-based** workers, firms and institutions to either defend existing jobs and economic investments or else attract new ones.

Labour supply
The specific quantity and quality of workers who make their **labour power** available for purchase by employers. Labour supply, like **labour demand**, is highly variable temporally, spatially and among firms and sectors.

Living wage
The amount of money needed to ensure the basic **physical** and **social reproduction** of a person and their family/dependents. Living wages vary with the cost of living.

Living wage campaigns
Organized grass-roots attempts to get states and firms to raise the wage minimum to a 'living wage' level.

Local
The relatively small **geographical scale** at which daily life is played out.

Locality
A term used in human geography and sociology to refer to the specific combination of people, infrastructure, industry, consumption practices, etc. that make up a **place.** A locality cannot be understood simply with reference to what happens at the local level. Instead, localities are understood to be the product of how local events *combine* with processes and relations operating over a broader **space**.

Local growth coalitions
Alliances between actors whose interests are distinctly **local** in **geographical scale.** Growth coalitions aim at boosting the economic wealth of a locality and can involve partnerships between local workers, local businesses and the local state, among others. They are sometimes know as 'cross-class alliances'.

Non-local scales
Geographical scales at which social relations, processes, institutions and events are organized that include but also transcend the local scale. Non-local scales are socially constructed and also influence the things they 'contain'.

Oligopoly
This term describes an economic sector dominated by a small number of large firms.

Physical reproduction
Both the biological reproduction of people (through childbirth) and the daily physiological reproduction of people (through eating, sleeping, etc.).

Place
A highly complex term that, at its simplest, refers to (i) the local scale and (ii) a distinct point on the earth's surface. Places have both 'objective' and 'subjective' dimensions. The former consists of the specific collection of people, buildings, environments and institutions that comprise a place. The latter involves the attachment to place people have as well as their place identities. In both cases, what makes a place in the contemporary world involves a combination of local happenings and non-local events that impact on that place (see **locality**).

Place-based
Any aspect of production, reproduction, consumption or regulation that occurs partly at the local scale but that is not necessarily **place-bound**.

Place-bound
Any aspect of production, reproduction, consumption or regulation that cannot 'escape' the local or sub-local scale. See **locality dependence**.

Power repertoires
The ensemble of strategies and tools available to given sets of workers and employers in given times and places. These strategies and tools are used to influence the terms and conditions of paid work and can be used in a consensual or conflictual manner.

Primary workers
A generic category describing those wage workers at the top of capitalist labour markets. That is those who are, variously, highly educated, highly paid, highly skilled, and highly employable.

Pseudo-commodity
A commodity that differs from inert commodities like tables or cars. A pseudo-commodity – like workers – differs from other commodities in two main respects. First, it is not permanently sold to a buyer. Rather, it is temporarily purchased. Second, because it is sentient it cannot be disposed of by its purchasers in any old way. Rather, its use must be negotiated between seller and buyer.

'Real' commodity
Any entity – whether physical (for example, a chair) or non-physical (for example, an idea) – that can be sold in return for money. All commodities have both a use-value and an exchange-value. Though all commodities are qualitatively unique, money provides a common measure that allows their value to be compared and their exchange to occur. Some anthropologists have taken a broad view of commodities that includes those in non-capitalist societies. In this book, though, we refer to capitalist commodities only. These are things that are produced with the intention of exchanging them for money and, for capitalists, making a profit.

'Reciprocal unionism'
The working of unions with the communities they are part of. The basic principles of reciprocal unionism are: (i) it allows unions to reach those

'hard to organize' workers who are not likely to be organized in the workplace; (ii) it allows unions to reach those workers who have felt excluded from trade unions; (iii) finding allies and building networks gives trade unions an important source of support for local workplace organizing campaigns; (iv) it provides a means of building alliances between public sector workers and service users; (v) it provides a mechanism to foster reciprocity between trade unions and community groups; (vi) such coalitions allow unions (and others) to take advantage of state strategies targeting localities for redevelopment and; (vii) community-union place-based coalitions might decide to provide new services.

Remittances
The income sent by migrants back to their places of origin or the places where their dependents live.

Secondary workers
A generic category describing those wage workers with relatively well-paid, secure jobs that demand a reasonable level of education, skill and training. Secondary workers are often 'blue' and 'white collar' workers and, numerically, comprise the majority of workers in capitalist economies.

Segmentation
A complex process whereby different workers (defined by age, gender, nationality, ethnicity, place, skill level, etc.) end up working in different occupations within and between firms and economic sectors. Segmentation occurs in labour demand, labour supply and workplace practices. It contributes to the production of different **class fractions** in capitalist societies.

Social division of labour
The division of jobs and production activities within the wider economy.

Social regulation
The complex process whereby social relationships between people are lent a certain order and regularity. Social regulation is both planned and unplanned, explicit and tacit. It involves set of institutions – from the state to the family – and sets of habits, norms and customs.

Social relations
A generic term used in this book to refer to any form of relationship – direct or indirect, local or non-local – between people. Social relations can be economic, cultural, political, financial or familial.

Social reproduction
The short-and long-term process whereby people undertake the non-biological and non-physiological practices necessary to sustain themselves. The process involves socialization, leisure, education, training, friendships, family relations and a raft of institutions, especially outside the workplace. Social reproduction involves the formation of identities, cultural norms and mores among individuals and groups. In Katz's (2001: 714) words, it 'entails acquiring and assimilating the shared knowledge, values and practices of the groups to which one belongs by birth or choice'.

Solidarity
Acts of mutual cooperation between, and loyalty among, workers at a variety of geographical scales. Solidarity is most difficult at the transnational scale between of the socio-geographic distance and difference between wage workers.

Space
A complex concept that, in this book, refers to the way certain socio-economic and other human relationships are 'stretched-out' over **geographical scales** that are more than simply **local**. These relationships both constitute space and are affected by it.

Spatial division of labour
The geographical expression of the **social division of labour**. Different places are characterized by different clusters of firms and workers.

Structures of feeling
A term coined by the cultural analyst Raymond Williams (1975). It refers to a shared sense of a place among a locality's residents. This sense emerges from the daily routines of life and from shared habits and customs.

Sunk costs
These are a form of costs that cannot be recovered once committed in a particular place. They have no real market value (though they have paper or hypothetical value), and cannot be recouped by selling a part of the plant, capital or equipment to a competitor (Clark and Wrigley, 1995). An appreciation of the concept can enrich our understandings of the geographies of investment, disinvestment, plant closure, and exit from particular industries. Most significant in this context is the insight that sunk costs may restrict the mobility of large, multi-locational

businesses. Strategic options such as firm exit and plant closure come to be seen as extreme decisions, most likely taken only after a variety of other possibilities have been exhausted, since the firm will suffer significant financial losses. Bizarrely, in some situations, it may actually be less damaging for a firm to incur losses in a particular place than to pay the costs of closure. In addition to acting as barriers to exit, sunk costs may also act as barriers to entry in certain industries, as when the need for high sunk costs may prove a disincentive to initial investment. But what exactly constitutes a sunk cost? Clark and Wrigley (1995) identify three types, each of which may relate to the embeddedness of production in place-specific labour market conditions: (i) there are 'set-up' sunk costs (for example, training of staff for a new facility); (ii) there are 'accumulated' sunk costs (the ongoing costs of doing business, for example maintaining business relationships) and; (iii) there are 'exit' sunk costs (for example, the pension entitlements of displaced labour).

Technical division of labour
The division of production tasks within a workplace and firm. When combined, the various tasks make one or more commodities that comprise the firms' product/s.

Tertiary workers
A generic category that refers to wage-workers on the fringes of capitalist labour markets, such as part-time workers, the low-paid and low-skilled, temporary workers and so on. Tertiary workers typically do not enjoy the rights and benefits of **primary** and **secondary** workers.

Time-space compression
The increase in the number of events happening in any given moment of real time and the reduction in the relative distance between places due to enhanced transport and communications technologies.

Trade unions
Trade unions are worker organizations created to protect and advance the interests of their members by negotiating agreements with employers on pay and conditions of work. Unions may also provide legal advice, financial assistance, sickness benefits and education facilities. In addition, they frequently lobby government and regulatory bodies on behalf of their membership. There are four key types of unions: (i) there are *craft* or *occupational* unions that represent workers doing similar tasks (for example, the UK's Professional Footballers Association); (ii) *employer-specific* unions are established to undertake

collective bargaining for workers with a single employer (for example, the UK's Alliance and Leicester Group Union of Staff – ALGUS); (iii) *industrial* unions exist where the aim is to organize workers in an entire industry without regard to occupational difference (for example, the UK's National Union of Mineworkers) and; (iv) with *general* unions the organizing pattern is truly collective rather than being based purely on occupation or industry (for example, the UK's Transport and General Workers Union). While these latter unions often have concentrations of members in particular occupations, they aim to represent all workers and are usually conglomerates of separate industrial unions. The distinctions between these different types of unions have become increasingly blurred due to waves of rationalization and merger that have occurred in response to falling levels of union membership across the industrialized world over the last few decades. In particular, more general unions have emerged as organizations and have come together in an attempt to preserve or enhance their bargaining power.

Translocalism
Any set of relations, events or processes that implicate two or more otherwise different and discrete places.

Transnationalism
Any set of relations, events or processes that implicate two or more otherwise different and discrete places in different countries. Transnational ties can thus be macro-regional (extending across a contiguous group of countries), international (extending over many countries, not necessarily contiguous ones) or global.

Transnational capital class
These are the relatively small number of powerful – and well-connected – individuals, working through institutions that they either own or control, who shape the global capitalist system. Leslie Sklair (2001: 1) terms this group as the 'transnational capitalist class' (TCC), which he suggests has 'transformed capitalism into a globalizing project'. The TCC is a part of the capitalist class of firm owners, but also includes regulators of workers and firms. It comprises four key components or 'fractions': (1) TNC executives and their local affiliates (the corporate fraction); (2) globalizing bureaucrats and politicians (the state fraction); (3) globalizing professionals (the technical fraction); and (4) merchants and the media (the consumerist fraction). The common characteristic is that these people operate *internationally* as a

normal part of their working lives, and it should be recognized that there is considerable overlap and mobility between the four groups identified above. Together the groups constitute a global power elite. The ongoing hegemony of the TCC does not happen by accident, rather 'the capitalist class expends much time, energy, and resources to make it happen and to ensure that it keeps on happening' (Sklair, 2001: 18).

Transnational communities
Defined by 'the formation and maintenance [across international boundaries] of "community", based especially on social, economic, and political networks, the construction and expression of identity focussed on the refashioning of cultural forms and symbols, and the reproduction or contestation of social relations including issues of gender and power' (Vertovec, 1999: 457). Transnational communities are communities with propinquity.

Use value
The practical benefits derived from the uses to which a thing (be it a potentially sellable item or not) can be put by virtue of its physical properties. In geographical terms, places yield multiple use values for those who live and work in them. For firms especially, these use values are directly linked to the question of **exchange values**.

Varieties of labour
The word 'labour' refers to both the act of work and the people who perform it (workers). Work and workers can be broadly divided into six categories: (i) the *self-employed* are just that: they work for themselves and their dependents; (ii) *household* or *domestic work* is similarly controlled by the worker or their family/co-dependents and involves no monetary payment; (iii) *volunteer work*, as the name suggests, is typically non-renumerated but, unlike domestic work, normally performed outside the home; (iv) *communal work* involves a community organizing its collective labour to achieve certain agreed ends; it can be paid or non-paid; (v) *non-monetized work* involves an individual, individuals or groups exchanging non-monetary goods in payment for work done (so-called 'payment in kind') and; (vi) *monetized work* involves an individual, individuals or groups exchanging money in payment for work done.

Wage work
Work undertaken by individuals and groups on behalf of others in return for money. In capitalist societies, wage work involves a class

relation between those who sell their **labour power** and those who purchase it.

Work

The activity of transforming physical or non-physical things into new products or else modifying, maintaining, moving, gathering or selling existing products. There are many categories of work undertaken in many different kinds of contexts.

Workplace

Workplaces are the physical locations where commodity production, sale and maintenance occur. The 'classic' workplace is the factory but today this is in no way representative of the variety of workplaces worldwide. We can distinguish workplaces by their physical form and, second, by their workers. In the first case, we can distinguish fixed site workplaces (for example, an assembly plant) from mobile workplaces (like an aircraft). In the latter case, we can distinguish (i) workplaces where all workers have the same employer, (ii) multi-employer workplaces (for example, an airport where airline employees mix with, say, security guards hired by a security firm contracted to the airport authority) and (iii) workplaces with a high fluidity of workers (as in those that employ temporary workers supplied by temp agencies). In the cases of (ii) and (iii) it is very hard for workers in the same workplace to achieve solidarity and mutual bonds because of their varied working arrangements. Together workplaces comprise **firms**.

appendix

Different Approaches to Theorizing Labour

As stated in the Preface, this book combines the Marxian and institutionalist approaches to understanding labour. Both dissent from the neo-classical view, which was once influential in the discipline of economics and elsewhere. In this short appendix we outline the characteristics of the three approaches in a basic rather than comprehensive way. For a more detailed discussion see pages 5–13 of Tilly and Tilly's (1998) *Work Under Capitalism.* Note that each 'approach' is, in reality, a cluster of related approaches rather than the homogenous entities we depict here. In this book, we combine facets of the Marxian and institutionalist approaches and are critical of the neo-classical worldview.

THE NEO-CLASSICAL APPROACH

The neo-classical approach emphasizes the free market as the key mechanism for equilibriating the supply of and demand for commodities. It sees market exchanges as free and equal transactions between individuals with certain needs and wants. So at point E_0 the

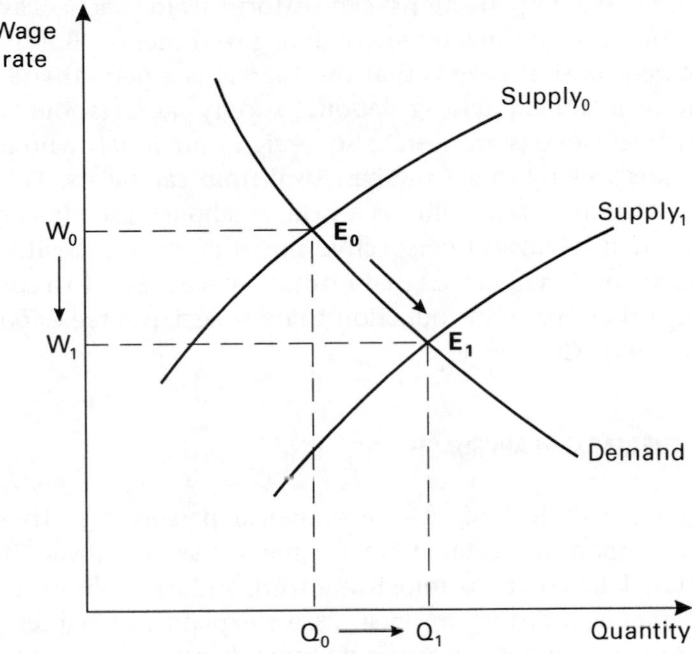

Appendix 1 The neo-classical model of labour supply and demand

market is in equlibrium. Supply is exactly the same as demand. If there is an increase in supply, as there is every summer for graduates, and demand remains unchanged then the wage rate falls from E_0 to E_1. The wage rate falls as the number of workers in the labour market increases. The buying and selling of labour is regarded as a commodity exchange like all others. Workers sell their services to purchasers (employers) and both parties get what they want – in theory at least. Thus employers get workers' capacity to labour for a finite period, while workers get wages that allow them to purchase the commodities they need to live and prosper. This does not mean that all workers are equal. On the contrary, the different skills and abilities of different workers ('human capital') mean that they are not 'worth' the same. However, whatever their individual market value, from the neo-classical perspective all workers should 'get what they deserve'. That is, they are paid what their skills warrant in an open market. In the neo-classical worldview, this market consists of millions of individuals engaging in monetary exchanges in a free, unencumbered way. Of course, 'market imperfections' can distort this in practice, as well the intervention of non-market forces (like governments). But the essence of the neo-classical view is that the market is a neutral and effective mechanism for matching labour supply and labour demand. Meanwhile, workers are seen as sovereign individuals who are 'rational' actors seeking to get the best deal from capitalists. Though this presupposition of rationality is a nod to labour's pseudo-commodity status, the neo-classical perspective nonetheless works with a highly unrealistic notion of labour that ultimately reduces it to a commodity like any other. For an introduction to neo-classicism see Caporaso and Levine (1991: Chapter 4).

THE MARXIAN APPROACH

Marxists are critical of the neo-classical perspective. They see its emphasis upon unregulated labour markets as unrealistic, its notion of 'rational' labourers sociologically thin, and its apolitical reading of the market as a distortion. First, as we explain in Chapter 2 of this book, labour markets can never be 'pure' because it is *people* who are being exchanged – albeit for certain daily periods of time only. For example, because workers might want to maximize their salaries, and because employers will want workers to be maximally productive, the buying and selling of workers is a potentially contradictory, tense affair. It is, in other words, a site of struggle not simply free exchange. Secondly, for Marxists the buying and selling of labour is deeply political, which is to say about social power and resistance. Capitalism is an inherently contradictory production system that generates unemployment as part of its 'logic'. What's more, Marxists regard workers as the origin of profit. In the so-called 'labour theory of value', Marxists suggest that the profits capitalists make derive ultimately from the fact that labour is paid less than the value of the goods it makes. This means that the neo-classical idea of the market as a neutral means of exchange hides the *exploitation* that goes on daily in the workplace. For Marxists, this exploitation pits workers against capitalists in both theory and practice. Contrary to the neo-classical vision of sovereign individuals buying and selling labour power, Marxists see a class politics where group interests condition individual interests.

Neo- and post-Marxists have sought to add to the basic theory of capitalist labour outlined above. First, since the early 1970s they've argued that many seemingly non-capitalist institutions, like the education system, are in fact geared to perpetuating class inequity such that workers remain ultimately subordinate to capitalists. Secondly, they've also debated whether workers are 'victims' of an insuperable capitalist system that offers them few choices but to work for others or whether they have 'agency'. In the latter case, attention has been focussed on how workers can resist or occasionally take the upper hand, namely from their employers. For more on Marxism see Swyngedouw (2000b).

THE INSTITUTIONALIST APPROACH

The institutionalist approach in economic geography emphasizes the role of institutions in economic life. Institutions are, according to Rutherford (1994, in Martin, 2000), 'a regularity of behaviours or a rule that is generally accepted by members of a social group, that specifies behaviour in specific situations, and that is either self-policed or policed by external authority' (p. 91). Such an approach can be contrasted with (non-institutionalist) neo-classical economics (where economic life is seen to depend on the behaviour of 'rational' individuals) or the Marxian approach (which focuses on class relations, has a more political stance against capitalism, and tends to be weaker on theorizing the role of institutions). Thus, in both neo-classical economics and to a lesser extent in the Marxian approach, economic life tends to be abstracted from its socio-political and cultural context. Thus, Martin (2000) defines the institutionalist approach in economic geography as responding to this basic question: 'to what extent and in what ways are the processes of geographically uneven capitalist economic development shaped and mediated by the institutional structures in and through which these processes take place?' (p. 79). In answering this question, institutionalist economic geography focuses on four issues. The first is the role of different sorts of institutions. In this sense, it is common to distinguish between the institutional environment (which are the rules, customs, social routines, and socialized work practices, etc.) and institutional arrangements (which refer to the organizational forms such as markets, firms, and labour unions, etc.) that both govern and arise from the institutional environment. Secondly, institutionalist approaches pay close attention to the evolution of the economic landscape (such as 'path dependence' and technological innovation). Thirdly, institutional economic geographers tend to focus on the cultural foundations of economic processes. That is the way in which attitudes, lifestyles, work traditions, union structures and so on, contribute to local and regional economic development. A fourth concern is the social regulation and governance of local and regional economies (such as welfare policies and employment or financial regulations).

But what implications does this have for the study of work and labour? Economic geographers interested in labour markets have taken a strong interest in the role of institutions, precisely because labour and employment are embedded in local, national, and macro-regional institutions. This gives local labour markets distinctive features that in turn both benefit and damage the livelihood of workers. Yet, it is not just institutions that shape the reproduction of labour and employment practices, but labour that shapes the institutional fabric of the space economy.

Further Reading and Key Questions

Below are a set of questions and readings relevant to each chapter of the book. All the references below can be found in the book's bibliography.

CHAPTER 2

Further reading

Below are listed essential follow-up readings for the various points made in this chapter:

- The overall argument in this chapter is given a longer and very accessible treatment in Tilly and Tilly (1998).
- A succinct explanation of what makes capitalism a distinctive mode of production can be found in Harvey (1985a: 128–33).
- Peck (1996: 24–6) elaborates on labour's 'pseudo-commodity' status and on worker-employer tensions (1995: Chapter 1).
- Gough (2001) deals nicely with the production–reproduction link.
- Hudson (2001: Chapters 5–7) is excellent on what divides employers and differentiates workers socially.
- Peck (1996: Chapter 3) offers a lucid explanation of segmentation.
- Mann (2001) uses a Los Angeles case study to show how different 'differences' among workers can be brought together productively in actions against firms and regulatory bodies.
- Finally, Storper and Walker (1989: Chapter 6) cover much of the same ground as this chapter, though put less of an emphasis on reproduction, regulation and 'non-capitalist differences'.

Key questions

Use this chapter and the further readings to answer the following questions:

- Why, in capitalist societies, do 'classes in-themselves' not readily become 'classes for-themselves'?
- What are some of the links and tensions between wage work and social reproduction?
- In what ways do social differences among wage-workers 'interfere' with their wider class consciousness?
- What difficulties do employers confront when trying to hire and subsequently control the 'right' workers?

CHAPTER 3

Further reading

- Two very accessible essays on how best to conceptualize places are those by Castree (2003) and Massey (1995a).
- For more on labour's localness see Hudson (2001: Chapter 8).
- Storper and Walker (1989: Chapter 5) provide a review of the territorial basis of economic production.
- Peck (1996: Chapter 4) explains why labour markets and their regulation are necessarily local and locally variable; see also Martin (2001) and Reimer (2001).
- The essay by Herod (1991) and the subsequent debate (Herod, 1994; Martin et al., 1994) explains very well how and why local identities clash with a wider inter-place class consciousness.

Key questions

- How can places remain unique at a time of heightened inter-place connectivity?
- For what reasons are the lives of wage-workers and related social groups necessarily local?
- Why must capitalist production and the regulation of firms and workers ultimately take place at the local scale?
- What are the implications of the 'localness' of labour for class identity and class action?

CHAPTER 4

Further reading

- It is well worth reading Massey's original account (1984: Chapter 3) to learn more about corporate spatial divisions of labour.
- Peck and Tickell (1995) show how local regulatory processes are embedded in wider institutional forms.

- For lucid introductions to geographical scale see Herod (2003) and Howitt (2003); the special issue of *Political Geography* (1997, number 2) provides a rich set of case studies of how scales are socially constructed; finally, Sheppard and McMaster's (2003) collection lucidly explains and illustrates the importance of geographical scale.
- See Swyngedouw (2000a) for a review of the contemporary re-scaling of the global economy.
- Jonas (1997) offers a clear account, with examples, of local labour control regimes.

Key questions

- What are the principal categories of inter-place connection and dependency linking capitalist firms and workers respectively?
- In what ways are corporate spatial divisions of labour changing, and how do these processes affect different groups of workers?
- In what ways do local labour control regimes necessarily reflect processes operating at wider scales?
- What kinds of scalar dilemmas do wage workers face in the contemporary world?

CHAPTER 5

Further reading

- Peck (1996: Chapter 8) provides a run through of the 'hegemonic despotism' thesis; Greider (1997: Chapters 1–6) paints a grim picture of workers and places competing ruthlessly against one another; Sparke and Lawson (2003) explain how nearby places can unite against rival places for jobs, using a North American case study; finally, Sheppard (2000) explains lucidly why inter-place competition is endemic to capitalism.
- Hudson (2000: Chapter 3) describes the ongoing problems in the economy and labour markets of Northeast England; Moody (1997: Chapter 5) provides an excellent review of the challenges to workers posed by lean production techniques across the developed world.
- Allen and Henry (1995, 1997) provide more detail on the 'downside' of growth in Southeast England; see Sassen (2001) for a detailed analysis of the highly segmented labour markets in London, New York and Tokyo.
- Kelly (2001, 2002) provides more detail on the Southeast Asia case study localities; see Cravey (1998) for a study of gendered industrialization in the Mexican *maquiladoras*.

Key questions

- To what extent is the notion of 'hegemonic despotism' useful in representing the fortunes of workers under global capitalism?
- What processes have contributed to the increased geographic mobility of capital?
- In what ways are business- and state-imposed mechanisms of labour control interrelated?
- What challenges does the 'down-scaling' of labour market regulation pose to workers and worker organizations?
- How and why do the mechanisms of labour control employed vary between different places?

CHAPTER 6

Further reading

- Cox and Mair (1988) illustrate how the concept of 'local dependence' illuminates our understanding of the politics of economic development in place.
- Herod (2001: Chapter 2) explains how and why wage-workers have agency; for an introduction to Giddens' idea of structuration see Layder (1994: Chapter 8).
- Logan and Molotch (1987: Chapter 3) provide the best definition of the 'growth machine' and explain why different interest groups are involved in place-based coalitions.
- Wills (2002) provides a good introduction to the concept of 'community unionism', with examples of differences in how it is defined in particular places, drawing on work in the United States and the UK.

Key questions

- Why do local firms, workers and state regulators sometimes mobilize together to defend their interests in place?
- How can labour work with others in the local community to assert its agency?
- What difficulties do unions face when trying to organize different local groups around a single issue in a particular place?
- What is the rationale for, and what are the limits to, organizing 'outside' of the global capitalist system?

CHAPTER 7

Further reading

- For an overview of the volume and diversity of migration flows, see Chapters 4, 5 and 6 in Castles and Miller (1998) and Chapter 6 in Held et al. (1999).
- All of the chapters in Brochmann and Hammar (1999) provide national case studies of the regulation and control of immigration in a selection of European countries. Chapter 5 in Jacobson (1998) provides a useful review of recent American immigration policies.
- Most of the chapters in the four-volume set (on Sub-Saharan Africa, South Asia, Latin America, and the Arab region) edited by Appleyard (1999) offer a wealth of excellent case studies of migration *within* these regions, and *to* other destination regions. They document the complexity of regulation, migration impacts, and the strategy of workers to improve their livelihoods.
- Sassen (1995) provides an excellent study of the impacts of migration at the scale of the local labour market, as does Borjas (in Jacobson, 1998) at both the national and the local scale in the United States.

Key questions

- To what extent does labour migration from countries in the Southern Hemisphere to the Northern Hemisphere constitute the bulk of labour migration?
- Explain why nation-states continue to be the most important regulators of labour migration.
- Why would states want to encourage the immigration of workers?
- In what ways does low-skilled labour migration affect the livelihoods of native workers?
- If professional workers are relatively affluent, why do they migrate?
- How, and to what extent, does the emigration of workers benefit source areas?

CHAPTER 8

Further reading

- Herod (2001: Chapter 5) offers a full account of up-scaling union activity from the local to the national scales, taking the longshoremen case summarized in this chapter as an example.

- Waterman and Wills' (2001) edited collection offers a range of excellent case studies on new initiatives in worker transnationalism.
- The essays by Herod (1995) and Castree (2000) try to pinpoint what can be learnt from recent transnational worker campaigns that are production-orientated. Are these campaigns unique or can more general lessons be drawn from them?
- Johns and Vural (2000) offer a lucid account of the Stop Sweatshops Campaign as one example of consumption politics; Hartwick (2000) makes a more general argument for consumption-orientated forms of worker activism; Smith (1996), using Starbucks coffee as an example, shows how commodity consumption can inadvertently perpetuate labour exploitation.
- Moody (1997) describes the contours of the new social movement unionism; Cohen and Rai (2000) discuss several new international grassroots movements, only some of which are strictly labour movements.

Key questions

- Why is it both desirable and increasingly possible for workers to organize themselves transnationally? What geographical contradictions must be negotiated in the process?
- What, if any, wider lessons can be learnt about 'successful' transnational worker actions from the Ravenswood and Liverpool cases?
- To what extent can consumption-oriented actions offer an alternative to production-focussed forms of worker transnationalism?
- Is borderless solidarity necessarily a good thing for contemporary wage-workers?

CHAPTER 9

Further reading

- Harvey (2001: Chapter 9) offers a readable account of the problem of reconciling different geographical scales at which wage workers define their loyalties to one another.
- Harvey (1996: Chapter 12) writes insightfully on the way the construction of a common rallying-point can help ameliorate 'negative' differences.
- On workers and human rights see Harvey (2000: Chapter 5) and Hensman (2001).
- Fraser (1997: Chapter 1) presents a lucid account of the tensions and complementarities that arise in the pursuit of economic and cultural justice, as does McDowell (2000); D.M. Smith (2000a and

b) offers a general reflection of how to achieve justice in a world of intense socio-geographic difference.

Key questions

- Is it possible to reconcile worker aims and objectives defined at one geographical scale with those defined at other geographical scales?
- How can identities, ideas or issues serve to protect 'positive' worker differences and eliminate 'negative' worker differences?
- Does moral relativism necessarily undermine attempts to improve the livelihoods of contemporary wage workers and their allies?

List of Websites and Other Learning Resources

WEBSITES

www.artsmonash.edu.au/projects/cep/index.htm
This website contains a series of projects looking at non-capitalist community economic development schemes in Australia.

www.labourstart.org/
This is a trade unionists' website that contains up-to-date stories about worker struggles worldwide.

www.laborart.com/
This site contains cartoons that satirically depict workers and their employers.

www.icftu.org/
The site of the International Federation of Free Trade Unions.

www.poptel.org.uk/women-ww/
A website devoted to reporting on women workers worldwide.

www.ilo.org/
The website of the International Labour Organization, a United Nations organization that monitors what's happening to workers worldwide.

www.labournet.org/
The International Labour Solidarity website. It's dedicated to sharing information about international worker campaigns to defend jobs, improve pay and enhance conditions.

theglobalalliance.org/
The Global Alliance for Workers and Communities website. This is an independent organization that reports worker rights abuses worldwide.

www.antenna.nl~waterman/
The Global Solidarity Dialogue website run by labour activist Peter Waterman. It's an information exchange site that encourages dialogue on new initiatives in transborder labour solidarity.

www.wftu.cz
The World Federation of Trade Unions website.
www.global-unions.org
The Global Unions website that reports on the activities of the several existing international union organizations.
www.labornotes.org
The home of Labour Notes, a regular report on labour activism worldwide.
www.unglobalcompact.org
Global Compact website, promoting social movement unionism that links worker struggles with those of communities, environmentalists and others.
www.global-labour.org
Website of the Global Labour Institute based in Geneva. This pro-worker organization is not tied to any one trade union and takes an ecumenical perspective on worker rights.
www.a16.org/
The Mobilization for Global Justice website. This site reports on activities that challenge the power of the WTO, the World Bank and the IMF.
www.citizen.org/trade/index.cfm
The Global Trade Watch website. This US-based site advocates socially and environmentally ethical trade among nations.
www.corpwatch.org/
The Corporate Watch website monitors the activities of multinational companies.
www.oneworld.net
OneWorld is a global internet based cooperative formed in 1995 and dedicated to harnessing the democratic potential of the internet to promote human rights and sustainable development.
www.tie.nl/
The Dutch-based Transnationals Information Exchange site. It also monitors multinational companies and their activities.
www.forumsocialmundial.org.br/home.asp
The website of the World Social Forum, a global organization of trade unions, new social movements, NGOs and other groups opposed to the neoliberal version of world capitalism.
www.caa.org.au/campaigns/nike/
This is a link off the Oxfam website that monitors Nike's employment practices.
www.etuc.org/
The European Trade Union Confederation was formed in 1973. It now has 78 national trade union confederations from 34 European countries.

It lobbies European politicians and decision-makers on behalf of its many millions of members.

www.etuc.org/etui

The European Trade Union Institute was formed in 1978 by the ETUC as a bridge between academic researchers and the trade union movement.

migration.ucdavis.edu

Migration News, a free leading source of monthly information on migration-related news across the world.

www.oecd.org

The Organization for Economic Cooperation and Development. See 'international migration' and especially their annual report *'Trends in International Migration'* published by their SOPEMI division.

www.homeoffice.gov.uk

The website of the United Kingdom Home Office, which contains a range of immigration matters relating to the UK and the European Union.

www.ercomer.org

The European Research Centre on Migration and Ethnic Relations – a general documentation centre containing a range of publications and notices.

www.nytimes.com

You can register for the *New York Times* online for free, and then click on the 'member center' and then 'news tracker' to receive daily notices about stories relating to immigration, migration, and so forth, in the United States and abroad.

VIDEOS AND FILMS

- 'Mickey Mouse Goes to Haiti'. This video is about the workers in Haiti who sew garments for Disney. Both videos are produced (and distributed) by the National Labor Committee in New York.
- 'Free Trade Slaves', by Joan Salvat, Stef Soetewey and Peter Breuls. Princeton, NJ: Films for the Humanities and Sciences, c. 1999. This film discusses free trade zones and the accompanying human problems that have arisen with human rights, exploitation of workers and environmental degradation. Filmed on location in Sri Lanka, El Salvador, Mexico and Morocco.
- 'Children at Work'. A documentary by Shelia Franklin. Amherst, MA: 1 World Producation, 2000. An examination of the use of child labour worldwide.
- 'Something to Hide'. National Labor Committee, New York: Crowing Rooster Arts, 1999. This film illustrates the long and

difficult hours that adults and children in developing countries are working to produce brand name American products.

- 'Secrets of Silicon Valley' Snitow-Kaufman Productions. Produced and directed by Alan Snitow and Deborah Kaufman, Bullfrog Films, 2001. This film chronicles a tumultuous year in the lives of two young activists grappling with rapid social change and the meaning of globalization on their own doorsteps. Magda Escobar runs Plugged In, a computer training centre in a low-income community just a few miles from the epicentre of high-tech wealth. Raj Jayadev is a temporary worker who confronts the hype of SiliconValley by revealing the reality of an unseen and unacknowledged army of immigrant workers.
- 'The Full Monty' (1997, Twenty-First Century Fox) and 'Brassed Off' (1997, Miramax Films). These two commercially successful films were shot in deindustrializing towns in Northern England. In different ways, they recount how the unemployed cope with being thrown out of work. 'Brassed Off' is altogether darker, and more tragic than the comedic 'The Full Monty'.
- 'Bread and Roses' (2001, FilmFour) (www.breadandroses. com) is Ken Loach's film based on the Justice for Janitors campaign in Los Angeles. It focuses on the politics of organizing low-paid, often illegal janitors in an effort to improve their lot.
- 'Real Women Have Curves' (2002, HBO/Newmarket Films) (www. realwomenhavecurves.com). This account of a Latino family from East Los Angeles reveals how sweatshop work is part and parcel of every day for millions of US immigrants. It explores the relationship between generations of women brought up to work long hours for low pay, and what happens when the youngest members of the family begin to want more from life.
- 'Modern Times' (1936, United Artists) (www.moderntimes. com). Modern Times is a story of the male worker and the dehumanizing effects of factory life. Extremely subversive at the time, Charlie Chaplin blends dry humour with a serious questioning of a critique of the way the industrial system squeezes the life out of individuals in the name of profit.

MAGAZINES AND JOURNALS

- *Red Pepper*: an independent red-green magazine that promotes socialist and ecocentric ideas. See http://www.redpepper.org.uk/
- *New Internationalist*: this magazine exposes injustice, poverty and inequality worldwide. See http://www.newint.org/

Bibliography

Aldridge, T., Lee, R., Leyshon, A., Thrift, N. and Williams, C.C. (1999) *Stroud LETs Report*, available at: www.geog.qmul.ac.uk//lets/reports.htm

Allen, J. and Hamnett, C. (1995) 'Uneven worlds', in J. Allen and C. Hamnett (eds), *A Shrinking World?* Oxford: Oxford University Press, pp. 233–54.

Allen, J. and Henry, N. (1995) 'Growth at the margins' in C. Hadjimichalis and D. Sadler (eds), *Europe at the Margins*, Chichester: Wiley, pp. 149–66.

Allen, J. and Henry, N. (1997) 'Ulrich Beck's *Risk Society* at work', *Transactions of the Institute of British Geographers* 22(2): 180–96.

Allen, J., Cochrane, A. and Massey, D. (1998) *Rethinking the Region*. London: Routledge.

Alvarez, M.R. and Butterfield, T.L. (1997) The resurgence of nativism in California? The case of Proposition 187 and illegal immigration, *Working Paper 1020, California Institute of Technology*, Division of Humanities and Social Sciences.

Amin, A. (1999) 'Placing globalization' in J. Bryson, N. Henry, D. Keeble and R. Martin (eds), *The Economic Geography Reader* Chichester: Wiley, pp. 40–5.

Amin, A. (2002) 'Spatialities of globalization', *Environment and Planning A*, 34 (3): 385–401.

Anderson, B. (1983) *Imagined Communities*, London: Verso.

Appleyard, R. (ed.) (1999) *Emigration Dynamics in Developing Countries* (4 Vols). Aldershot: Ashgate.

Auster, C. (1996) *The Sociology of Work: Concepts and Cases*. Pine Thousand Oaks, CA: Pine Forge Press.

Babbage, C. (1835) *On the Economy of Machinery and Manufacturers*. London: Charles Knight, Pall Mall East.

Bailey, A., Wright, R.A., Mountz, A. and Miyares, I.M. (2002) '(Re)producing Salvadoran transnational geographies', *Annals of the Association of American Geographers*, 92 (1): 125–44.

Bales, K. (1999) *Disposable People*. Berkeley: University of California Press.

Barke, M., Fuller, D., Gough, J., MacFarlane, R. and Mowl, G. (eds) (2001) *Introducing Social Geographies*. London: Arnold, pp. 13–41.

Bauder, H. (2003) 'Brain abuse: the devaluation of immigrant labour in Canada', *Antipode*, 35 (4): forthcoming.

Beaverstock, J. and Smith, J. (1996) 'Lending jobs to global cities: skilled international labour migration, investment banking and the city of London', *Urban Studies*, 33 (8): 1377–94.

Beaverstock, J. and Boardwell, J. (2000) 'Negotiating globalization, transnational corporations and global city financial centres in transient migration studies', *Applied Geography*, 20, 277–304.

Benner, C. and Rosner (1998) *Living Wage: An Opportunity for San José* (available at: http://www.wpusa.org/publications/sjLvgWage.htm).

Benner, C. (2002) *Work in the New Economy: Flexible Labour Markets*. Oxford: Blackwell.

Berger, J. (1974) *The Look of Things*. London: Viking Press.

Bernstein, H. (2001) 'The peasantry in global capitalism', in L. Panitch and C. Leys, pp. 25–52.

Berriane, M. and Popp, H. (eds) (1998) *Migrations internationales entre le Maghreb et l'Europe*. Passau: L.I.S. Verlag.

Beynon, H. and Hudson, R. (1993) 'Space and place in contemporary Europe', *Antipode*, 25 (3): 177–90.

Betcherman, G. (1996) 'Globalization, labour markets and public policy', in R. Boyer and D. Drache (eds) *States against Markets: The Limits of Globalization*. London: Routledge, pp. 250–69.

BIS (2001) *71st Annual Report*, Bank for International Settlements, Basle.

Bolt, P.J. (2000) *China and Southeast Asia's Ethnic Chinese*. London: Praeger.

Bonacich, E. (2001) 'The challenge of organizing in a globalized/flexible industry: the case of the apparel industry in Los Angeles in C. Baldoz, C. Koeber and P. Kraft (eds), *The Critical Study of Work: Labor, Technology and Global Production*. Philadelphia: Temple University Press pp. 155–78.

Borjas, G.J. (1999) *Heaven's Door: Immigration Policy and the American Economy*. Princeton: Princeton University Press.

Borrus, M. (2000) 'The resurgence of US electronics', in M. Borrus, D. Ernst and S. Haggard (eds) *International Production Networks in Asia*. London: Routledge, pp. 57–79.

Bradshaw, M.J. (1996) 'The prospects for the post-socialist economies', in P.W. Daniels and W.F. Lever (eds) *The Global Economy in Transition*. Harlow: Longman, pp. 263–88.

Braverman, H. (1974) *Labour and Monopoly Capital*. London: Monthly Review Press.

Brecher, J. and Costello, T. (1998) *Global Village or Global Pillage* (2nd edn), Boston: South End Press.

Brochmann, G. and Hammar, T. (1999) (eds) *Mechanisms of Immigration Control*. Oxford: Berg Press.

Burawoy, M. (1985) *The Politics of Production*. London: Verso.

Cabinet Office (2002) *Interim Report on Ethnicity and Work in the United Kingdom*. London: HMSO.

Caporaso, J. and Levine, D. (1991) *Theories of Political Economy*. Cambridge: Cambridge University Press.

Castells, M. (1997) *The Power of Identity*. Oxford: Blackwell.

Castles, S. and Miller, M. (1998, 2nd ed) *The Age of Migration: International Population Movements in the Modern World*. London: Macmillan.

Castree, N. (2000) 'Geographical scale and grassroots internationalism', *Economic Geography* 76 (3): 272–92.

Castree, N. (2003) 'Place', in S. Holloway, S.P. Rice and G. Valantine (eds) *Key Concepts in Geography*. London: Sage pp. 252–82.

Clamp, C. (2000) 'The internationalization of Mondragón', *Annals of Public and Cooperative Economics*, 71, 557–77.

Clark, G.L. and Wrigley, N. (1995) 'Sunk costs: a framework for economic geography', *Transactions of the Institute of British Geographers*, 20 (2): 204–23.

Cockburn, C. (1977) *The Local State: Management of Cities and People*. London: Pluto Press.

Cochrane, A. (1993) *Whatever Happened to Local Government?* Milton Keynes: Open University Press.

Coe, N.M. and Kelly, P.F. (2000) 'Distance and discourse in the local labour market', *Area* 32: 413–22.

Coe, N.M. and Townsend, A.R. (1998) 'Debunking the myth of localized agglomerations', *Transactions of the Institute of British Geographers*, 23 (3): 385–404.

Coe, N.M. and Yeung, H. (2001) 'Geographical perspectives on mapping globalization', *Journal of Economic Geography*, 1 (4): 367–80.

Cohen, R. and Rai, S. (2000) *Global Social Movements*. London: Athlone Press.

Connor, T. (2002) 'We are not machines', *Report for the Clean Clothes Campaign*, Amsterdam, Clean Clothes Campaign.

Cox, K.R. (1993) 'The local and the global in the new urban politics: a critical view', *Environment and Planning D: Society and Space*, 11, 433–48.

Cox, K.R. (1998) 'Spaces of dependence, spaces of engagement and the politics of scale', *Political Geography*, 17 (1): 1–24.

Cox, K.R. and Mair, A. (1988) 'Locality and community in the politics of local economic development', *Annals of the Association of American Geographers*, 78, 307–325.

Cox K.R. and Mair (1991) 'From localised social structures to localities as agents', *Environment and Planning A*, 23, 197–213.

Crang, P. (1999) 'Local-global', in P. Cloke, P. Crang and M. Goodwin (eds), *Introducing Human Geography*. London: Arnold, pp. 24–34.

Crang, P. and Martin, R.L. (1991) 'Mrs Thatcher's vision of the "new Britain" and the other sides of the "Cambridge phenomenon"', *Environment and Planning D*, 9: 91–116.

Cravey, A. (1998) *Women and work in Mexico's Maquiladoras*, Oxford: Rowman and Littlefield.

Cresswell, T. (1999) 'Place', in P. Cloke, P. Crang and M. Goodwin (eds), *Introducing Human Geography*. London: Arnold, pp. 226–34.

Dawley, S. and Pike, A. (2001) 'The ebb and flow of TNC high technology (dis)investment in host region economies'. Paper presented to the 97th Annual Meeting of the Association of American Geographers, New York.

Delauney, D. and Tapinos, G. (1998) *La mesure de la migration clandestine en Europe: Volume Rapport de Synthèse*. EUROSTAT Working Paper 3/1998/E/no.7 (EUROSTAT).

Dicken, P. (1998) *Global Shift* (3rd edn). London: Paul Chapman.

Dicken, P. (2003) *Global Shift* (4th edn). London: Sage.

Dicken, P., Peck, J. and Tickell, A. (1997) 'Unpacking the global', in R. Lee and J. Wills (eds) *Geographies of Economies*. London: Arnold, pp. 158–66.

Dicken, P., Kelly, P., Olds, K. and Yeung, H.W-C. (2001) 'Chains and networks, territories and scales', *Global Networks* 1: 89–112.

The Economist, 'South Africa's migrant workers: a ticket to prosperity', 2 September 2000.

The Economist, 'Desperate cargo: migrants want to fight human trafficking and smuggling in Asia', 3 February 2002.

The Economist, 'Troubled waters', 13 July 2002.

Edwards, P. and Elger, T. (1999) (eds), *The Global Economy, National States and the Regulation of Labour*. Mansell.

Elson, D. and Pearson, R. (1984) 'The subordination of women and the internationalization of factory production', in K. Young, C. Wolkowitz and R. McCullagh (eds) *Of Marriage and The Market*. London: Routledge, pp. 18–40.

Enderwick, P. (1989) 'Multinational corporate restructuring and international competitiveness', *California Management Review*, 32: 44–58.

Fagan, R.H. and Webber, M. (1999) *Global Restructuring*: *The Australian Experience* (2nd edn). Melbourne: Oxford University Press.

Fan, C.C. (2002) 'The elite, the natives, and the outsiders: migration and labor market segmentation in urban China', *Annals of the Association of American Geographers*, 92 (1): 103–24.

Fitzgerald, J. (1991) 'Class as community: the new dynamics of social change', *Environment and Planning D*: *Society and Space*, 9: 117–28.

Fraser, N. (1997) *Justice Interruptus*, New York: Routledge.

Frey, W. (1996) 'Immigration, domestic migration, and demographic balkanization in America: new evidence for the 1990s', *Population and Development Review*, 22 (4): 741–63.

Fröbel, F., Heinrichs, J. and Kreye, O. (1980) *The New International Division of Labour*. Cambridge: Cambridge University Press.

Galbraith, J.K. (1995) 'A global living wage', in C. Crouch and D Marquand (eds), *Reinventing Collective Action*. Oxford: Blackwell, pp. 54–60.

Gallin, D. (2001) 'Propositions on trade unions and informal employment', in P. Waterman and J. Wills, pp. 531–49.

Geddes, A. (2000) *Immigration and European Integration*. Manchester: University of Manchester Press.

Gereffi, G. and Korzeniewicz, M. (1994) (eds) *Commodity Chains and Global Capitalism*. Westport, CT: Praeger.

Gibson-Graham, J-K. (1996) *The End of Capitalism (As We Knew It)*. Oxford: Blackwell.

Giddens, A. (1979) *Central Problems in Social Theory*. Berkeley: University of California Press.

Giddens, A. (1981) *A Contemporary Critique of Historical Materialism*. Berkeley: University of California Press.

Giddens, A. (1984) *The Constitution of Society*. Cambridge: Polity Press.

Giddens, A. (1990) *The Consequences of Modernity*. Cambridge: Polity Press.

Gills, B. (1997) 'Globalization and the politics of resistance', *New Political Economy*, 2 (1): 11–16.

GMB (2002) 'Where do you want your taxes to go?', *The Guardian* 1 April.

Gordon, D.M. (1996) *Fat and Mean*. New York: The Free Press.

Gough, J. (2001) 'Work, class and social life', in R. Pain, M. Barke, D. Fuller, J. Gough, R. Macfarlane and G. Mowl, *Introducing Social Geographies*. London: Arnold, pp. 13–42.

Gourevitch, P., Bohn, R. and McKendrick, D. (2000) 'Globalization of production', *World Development*, 28 (2): 301–17.

Gow, D. (2002) 'Where the grass is greener', *The Guardian*, June 13, page 7.

Greider, W. (1997) *One World, Ready or Not*. Harmondsworth: Penguin.

Guardian, The 'US state drops "racist" measure', 30 July 1999.

Guardian, The 'Death in the pursuit of the American dream', 15 January 2000.

Guardian, The (2001a) 'Delhi calling', G2, 9 March: 2–3.

Guardian, The (2001b) 'Motorola refuses to relent', 25 April: 23.

Guardian, The (2002a), 'Dyson turns to the Far East', 6 February: 3.

Guardian, The (2002b) 'End of the line', G2, 19 February: 3–5.

Guardian, The (2002c) 'Reading the numbers all right', 6 April: 30.

Guardian, The (2002d) 'The tale of a teabag', G2, 25 June: 1–5.

Habermas, J. (1976) *Legitimation Crisis*. London: Heinemann.

Hale, A. and Shaw, L. (2001) 'Women workers and the promise of ethical trade in the globalized garment industry', *Antipode* 33, (3): 510–31.

Hanson, S. and Pratt, G. (1995) *Gender, Work and Space*. London: Routledge.

Harrison, B. (1997) *Lean and Mean* (2nd edn). New York: Guilford Press.

Hartwick, E. (1998) 'Geographies of consumption', Environment and Planning D. *Society and Space*, 16: 423–37.

Hartwick, E. (2000) 'Towards a geographical politics of consumption', Environment and Planning A. *Society and Space*, 32: 1177–92.

Harvey, D. (1973) *Social justice and the city*. London: Arnold.

Harvey, D. (1985a) 'The geopolitics of capitalism', in D. Gregory and J. Urry (eds), *Social Relations and Spatial Structures*. London: Macmillan, pp. 128–63.

Harvey, D. (1985b) *The Urbanisation of Capital*. Oxford: Blackwell.

Harvey, D. (1989a) *The Urban Experience*. Oxford: Blackwell.

Harvey, D. (1989b) *The Condition of Postmodernity*. Oxford: Blackwell.

Harvey, D. (1993) 'From space to place and back again: reflections on the condition of postmodernity', in J. Bird, B. Curtis, T. Putnam, G. Robertson and L. Tickner (eds) *Mapping the Futures: Local Cultures, Global Change*. (London: Routledge), pp. 3–29.

Harvey, D. (1996) *Justice, Nature and the Geography of Difference*. Oxford: Blackwell.

Harvey, D. (2000) *Spaces of Hope*, Edinburgh: Edinburgh University Press.

Harvey, D. (2001) *Spaces of Capital*. Edinburgh: Edinburgh University Press.

Hayter, T. (1997) *The Dynamics of Industrial Location*. Chichester: Wiley.

Held, D. (1995) *Democracy and the Global Order*. Cambridge: Polity Press.

Held, D., McGrew, A., Goldblatt, D. and Perraton, J. (1999) *Global Transformations*. Polity Cambridge: Press.

Hensman, R. (2001) 'World trade and workers' rights', in P. Waterman and J. Wills, pp. 102–22.

Herod, A. (1991) 'Local political practice in response to a manufacturing plant closure', *Antipode*, 23 (4): 385–402.

Herod, A. (1994) 'Further reflections on organized labor', *Antipode*, 26 (1): 77–95.

Herod, A. (1995) 'The practice of international labor solidarity and the geography of the world economy', *Economic Geography*, 71 (4): 341–63.

Herod, A. (1997a) From a geography of labor to a labour geography: labor's spatial fix and the geography of capitalism, *Antipode*, 29: 1–31.

Herod, A. (1997b) 'Labor's spatial praxis and the geography of contract bargaining in the US east coast longshore industry, 1953–89', *Political Geography*, 16 (2): 145–69.

Herod, A. (1998) (ed.) *Organizing the Landscape*. Minneapolis: Minnesota University Press.

Herod, A. (2000a) 'Labor unions and economic geography', in E. Sheppard and T. Barnes (eds) *A Companion to Economic Geography*. Oxford: Blackwell, pp. 341–58.

Herod, A. (2000b) 'Workers and workplaces in the neoliberal economy', *Environment and Planning A*, 32: 1781–90.

Herod A (2000c) Implications of just-in-time production for union strategy: lessons from the 1998 General Motors – United Auto Workers dispute, *Annals of the American Geographers* 90: 521–47.

Herod, A. (2001) *Labor Geographies*. New York: Guilford Press.

Herod, A. (2003) 'Scale: the local and the global', in S. Holloway et al. (eds) *Key Concepts in Geography*. (London: Sage), pp. 229–249.

Herod, A., Tuathail, G.Ó. and Roberts, S.M. (1998) (eds) *An Unruly World?* London: Routledge.

Hiebert, D. (2002) 'The spatial limits to entrepreneurship: immigrant entrepreneurs in Canada', *Tijdschrift voor Economische en Sociale Geografie*, 93 (2): 173–90.

Hirst, P. and Thompson, G. (1999) *Globalization in Question* (2nd edn). Cambridge: Polity Press.

Hobsbawm, E. (2002) *Interesting Times*. Allen Lane, London.

Holloway, L. and Hubbard, P. (2000) *People and Place*. Prentice Hall: London.

Holloway, R. (2002) 'How to be good', *The Guardian*, supplement, 21 September.

Howitt, R. (2003) 'Scale' in J. Agnew, K. Mitchell and G. Toal (eds) *A Companion to Political Geography*. Oxford: Blackwell, pp. 138–57.

Hudson, R. (1989) 'Labour market changes and new forms of work in old industrial regions', *Environment and Planning D*, 7: 5–30.

Hudson, R. (2000) *Production, Places and Environment.* Prentice Hall: Harlow.

Hudson, R. (2001) *Producing Places.* New York: Guilford Press.

Hudson, R. and Sadler, D. (1986) 'Contesting work closures in Western Europe's old industrial regions', in A.J. Scott and M. Storper (eds) *Production, Territory, Work.* Allen and London: Unwin, pp. 172–93.

Hughes, A. (2000) 'Retailers, knowledge and changing commodity networks: the case of the cut flower trade', *Geoforum*, 31: 175–190.

Hyman, R. (1999) 'Imagined solidarities', in P. Leisink (ed), *Globalization and Labour Relations.* Cheltenham: Edward Elgar, pp. 94–115.

ILO (2001) *World of Work.* Geneva: ILO.

ILO Bureau of Statistics (2003) *Statistics of Trade Union Membership.* Geneva: ILO.

IOM (International Organization for Migration) *IOM News*, September 2001.

Jacobson, D. (1998) *The Immigration Reader.* Oxford: Blackwell.

Jakarta Post, 'Migrant workers: exporting people at risk', 10 March 2002.

Jessop, B. (1993) 'Towards a Schumpeterian workfare state? Preliminary remarks on post-Fordist political economy', *Studies in Political Economy* 40: 7–39.

Jessop, B. (2001) 'Multinational enterprises, national states and labour relations after the end of globalization', *International Journal of Urban and Regional Research* 25: 439–44.

Johns, R. and Vural, L. (2000) 'Class, geography and the consumerist turn. *Environment and Planning A*, 32: 1193–1214.

Johnston, R.J. Gregory, D., Pratt, G. and Watts, M. (2000) (eds) *The Dictionary of Human Geography* (4th edn). Oxford: Blackwell.

Jonas, A.E.G. (1992) 'Corporate takeover and the politics of community: the case of the Norton company in Worcester', *Economic Geography*, 68: 348–72.

Jonas, A.E.G. (1996) 'Local labour control regimes', *Regional Studies*, 30: 323–38.

Jonas, A.E.G. (1997) 'Localisation and globalisation tendencies in the social control and regulation of labour', in M. Taylor and S. Conti (eds) *Interdependent and Uneven Development.* Aldershot: Ashgate pp. 253–82.

Kang, N-H. and Sakai, K. (2000) 'International strategic alliances', *Directorate of Science, Technology and Industry Working Paper 2000/5*, OECD, Paris.

Katz, C. (2001) 'Vagabond capitalism and the necessity of social reproduction' *Antipode* 33, 4: 709–28.

Kelly, P.F. (2001) 'The political economy of local labor control in the Philippines', *Economic Geography*, 77 (1): 1–22.

Kelly, P.F. (2002) 'Spaces of labour control: comparative perspectives from Southeast Asia', *Transactions of the Institute of British Geographers*, 27(4): 395–411.

King, R. (1995) 'Migrations, globalization and place', in D. Massey and P. Jess (eds) *A Place in the World*? Milton Keynes: Open University Press.

Kloosterman, R., van der Leun, J. and Rath, J. (1999) 'Mixed embeddedness: (in)formal economic activities and immigrant businesses in the Netherlands', *International Journal of Urban and Regional Research*, 23 (2): 252–66.

Klotz, A. (2000) 'Migration after apartheid: deracializing South African foreign policy', *Third World Quarterly*, 21 (5):831–47.

Layder,D. (1994) *Understanding Social Theory*. London: Sage.

Lee, E. (1999) 'Trade unions, computer communications and the New World Order', in Munck and Waterman, pp. 229–47.

Leyshon, A. and Lee, R. (2003) 'Introduction', in R. Lee, A. Leyshon, and C. Williams (eds) *Alternative economic spaces*. London: Sage, pp 1–33.

Logan, J.R. (ed.) (2002) *The New Chinese City*. Oxford: Blackwell.

Logan, J. and Molotch, H. (1987) *Urban Fortunes: The Political Economy of Place*. (Berkeley: University of California Press).

Logue, J. (1980) *Toward a Theory of Trade Union Internationalism*. London: Allen & Unwin.

Mann, E. (2001) 'A race struggle, a class struggle, a women's struggle all at once', in L. Panitch and C. Leys, pp. 259–74.

Markusen, A. (1996) 'Sticky places in slippery spaces: a typology of industrial districts', *Economic Geography*, 72: 293–313.

Martin, R. (2000) 'Institutional approaches in economic geography', in E. Sheppard, and T. Barnes (eds) *A Companion to Economic Geography*. Oxford: Blackwell, pp. 77–94.

Martin, R.L. (2001) 'Local labour markets', in G. Clark, M. Feldman and M. Gertler (eds) *The Oxford Handbook of Economic Geography*. Oxford: Oxford University Press, pp. 455–76

Martin, R., Sunley, P. and Wills, J. (1994) 'Unions and the politics of deindustrialization', *Antipode* 26 (1): 59–76.

Martin, R.L., Sunley, P. and Wills, J. (1996) *Union Retreat and the Regions*. London: Jessica Kingsley.

Marx, K. and Engels, F. (1848) [2000] *A Communist Manifesto*. London: Penguin.

Marx, K. (1857) [1972] *Grundrisse*, in *Marx-Engels Collected Works*. London: Lawrence and Wishart.

Massey, D. (1984) *Spatial Divisions of Labour*. Basingstoke: Macmillan.

Massey, D. (1985) 'New directions in space, in D. Gregory and J. Urry (eds) *Social Relations and Spatial Structures*. London: Macmillan, pp. 9–19.

Massey, D. (1995a) 'The conceptualization of place', in D. Massey and P. Jess (eds), *A place in the world?* Oxford: Oxford University Press, pp. 46–79.

Massey, D. (1995b) 'Masculinity, dualisms and high technology', *Transactions of the Institute of British Geographers* 20: 487–99.

Massey, D. (1995c) *Spatial Divisions of Labour* (2nd edn). Basingstoke: Macmillan.

Massey, D. (1999) *Power-Geometries and the Politics of Space-Time*. Heidelberg: University of Heidelberg Press.

McDowell, L. (1999) *Gender, Identity and Place: Understanding Feminist Geographies*. Cambridge: Polity Press.

McDowell, L. (2000) 'Economy, culture, difference and justice', in I. Cook, D. Crouch, S. Naylor and J.R. Ryan (eds) *Cultural turns/geographical turns*. Harlow: Prentice Hall, pp. 182–95.

McDowell, L. (2001) 'Linking scales: or how research about gender and organizations raises new issues for economic geography', *Journal of Economic Geography*, 1: 227–50.

McDowell, L. (2003) 'Cultures of labour', in K. Anderson, M. Domosm, S. Pile and N. Thrift (eds) *Handbook of Cultural Geography*. London: Sage, pp. 98–115.

Meegan, R. (1995) 'Local worlds', in J. Allen and D. Massey (eds) *Geographical Worlds*. Oxford University Press, pp. 53–104.

Merrifield, A. (1993) 'Space and place', *Transactions of the Institute of British Geographers*, 18: 516–31.

Michon, F. (1987) 'Segmentation, employment structures and productive structures'. In R. Tarling (ed.) *Flexibility in Labour Markets*. London: Academic Press, 23–55.

Mitchell, D. (2000) *Cultural Geography: A Critical Introduction*. Oxford: Blackwell.

Mittelman, J.H. (2000) *The Globalization Syndrome: Transformation and Resistance*, Princeton, NJ: Princeton University Press.

Moody, K. (1997) *Workers in a Lean World*. London: Verso.

Mullings, B. (1999) 'Sides of the same coin?', *Annals of the Association of American Geographers* 89: 290–311.

Munck, R. and Waterman, P. (1999) (eds) *Labour Worldwide in an Era of Globalization*. London: Macmillan.

Natter, W. (1995) 'Radical democracy: hegemony, reason, time and space', *Society and Space*, 13 (3): pp 267–74.

New York Times, 'Migrant exodus bleeds Mexico's Heartland', 17 June 2001.

New York Times, 'A fertile farm region pays its jobless to quit California', 18 June 2001.

Nissen, B. and Grenier, G. (2001) 'Union responses to mass immigration: the case of Miami, USA', *Antipode*, 33 (3): 567–92.

O'Connor, J. (1973) *The Fiscal Crisis of the State*. London: St Martin's Press.

Ohmae, K. (1990) *The Borderless World*. New York: Free Press.

Ohmae, K. (1995) *The End of the Nation State*. Cambridge, MA: Harvard Business School.

Oonk, G. (2001) *The Dark Side of Football*. Amsterdam: The Netherlands India Committee.

Osterman, P. (1999) *Securing Prosperity: the American Labour Market How it has Changed and What to Do About It*. Princeton: Princeton University Press.

Outhwaite, W. and Bottomore, T. (1994) (eds) *The Blackwell Dictionary of Social Thought*. Oxford: Blackwell.

Panitch, L. and Leys, C. (2001) (eds) *Working Classes, Global Realities*. Merlin Press.

Pearson, R. (1993) 'Gender and new technology in the Caribbean', in J. Momsen (ed.) *Women and Change in the Caribbean*. Kingston, Jamaica: Ian Randle pp. 287–95.

Peck, J. (1996) *Workplace: The Social Regulation of Labor Markets*. New York: Guilford Press.

Peck, J. (2000) *Workfare States*. New York: Guilford Press.

Peck, J. and Theodore, N. (2001) 'Contingent Chicago: restructuring the spaces of temporary labor', *International-Journal-of-Urban-and-Regional-Research*, 25: 471–496.

Peck, J. and Tickell, A. (1992) 'Local modes of social regulation? Regulation theory, Thatcherism and uneven development', *Geoforum* 23: 347–63.

Peck, J. and Tickell, A. (1995) 'The social regulation of uneven development', *Environment and Planning A*, 27: 15–40.

Peet, R. (1998) *Modern Geographical Thought*. Oxford: Blackwell.

Pellerin, H. (1999) 'Regionalization of migration policies and its limits: Europe and North America compared', *Third World Quarterly*, 20 (5): 995–1011.

Picchio, A. (1992) *Social Reproduction*. Cambridge: Cambridge University Press.

Pickles, J. and Smith, A. (1998) (eds) *Theorizing Transition: The Political Economy of Post-communist Transformations*. London: Routledge.

Piven, F. and Cloward, R. (2000) 'Power repertoires and globalization', *Politics and Society* 28: 413–30.

Polanyi, K. (1944) *The Great Transformation*. Boston: Beacon Press.

Pollert, A. (1981) *Girls, Wives, Factory Lives*. Basingstoke: Macmillan.

Pratt, G. and Hanson, S. (1994) 'Geography and the construction of difference', *Gender, Place and Culture*, 1 (1): 5–29.

Raghuram, P. and Kofman, E. (2002) 'The state, skilled labour markets, and immigration: the case of doctors in England', *Environment and Planning A*, 34 (11): 2071–89.

Ram, M. (1998) 'Enterprise support and ethnic minority firms', *Journal of Ethnic and Migration Studies* 24 (1): 143–58.

Reimer, S. (2001) 'Conceptualising local labour markets', in Ron Martin and Philip Morrison (eds), *Local Labour Markets*. London: TSO, in press.

Rifkin, J. (1995) *The End of Work*. New York: Tarcher.

Roy, A.S. (1997) 'Job displacement effects of Canadian immigrants by country of origin and occupation', *International Migration Review*, 31 (1): 150–61.

Rutherford, T. and Gertler, M. (2002) 'Labour in "lean" times', *Transactions of the Institute of British Geographers*, 27 (2): 195–212.

Sadler, D. (2000) 'Organizing European labour', *Transactions of the Institute of British Geographers*, 25 (2): 135–52.

Said, E. (1978) *Orientalism*. New York: Pantheon Books.

Salt, J. (2000) 'Trafficking and human smuggling: a European perspective', in IOM (ed.) *Perspectives on Trafficking of Migrants*. Geneva: UN/IOM.

Salt, J. and Stein, J. (1997) 'Migration as a business: the case of trafficking', *International Migration*, 35 (4): 67–94.

Samers, M. (1998) '"Structured coherence": immigration, racism, and production in the Paris car industry', *European Planning Studies*, 6 (1): 49–72.

Samers, M. (1999) 'Globalization', migration, and the geo-political economy of the 'spatial vent', *Review of International Political Economy*, 6 (2): 166–99.

Samers, M. (2001) 'Here to work', *SAIS Review* Winter-Spring, XXI (1): 131–45.

Samers, M. (2003) 'Invisible capitalism: the political economy and regulation of undocumented immigration in France', *Economy and Society*, forthcoming.

Sassen, S. (1991) *The Global City*. Princeton, NJ: Princeton University Press.

Sassen, S. (1995) 'Immigration and local labor markets', in A. Portes (ed.), *The Economic Sociology of Immigration*. New York: Russell Sage.

Sassen, S. (2000) *Cities in a World Economy* (2nd edn). Thousand Oaks, CA: Pine Forge Press.

Sassen, S. (2001) *The Global City* (2nd edn). Princeton, NJ: Princeton University Press.

Saxenian, A (1994) *Regional Advantage*. Cambridge, MA: Harvard University Press.

Sayer, A. (1995) *Radical Political Economy*. Oxford: Blackwell.

Sayer, A. (2000) *Realism and Social Science*. London: Sage.

Sayer, A. and Walker, R. (1992) *The New Social Economy*. Oxford: Blackwell.

Schoenberger, E. (1997) *The Cultural Crisis of the Firm*. Oxford: Blackwell.

Schoenberger, E. (1998) 'Discourse and practice in human geography', *Progress in Human Geography*, 22: 1–14.

Scott, A.J. (1988) *New Industrial Spaces*. London: Pion.

Sharp, J.P., Ramledge, P. Philo. C. and Paddison, R. (eds) (2000) *Entanglements of Power*. London: Routledge.

Sheppard, E. (2000) 'Competition in space and between places', in E. Sheppard and T. Barnes (eds) *A Companion to Economic Geography*. Oxford: Blackwell, pp. 169–86.

Sheppard, E. and Porter, P. (1998) *A World of Difference: Society, Nature and Development*. London: Guildford Press.

Sheppard, E. and McMaster, R. (eds) (2003) *Scale and Geographic Inquiry*. Oxford: Blackwell.

Sklair, L. (2001) *The Transnational Capitalist Class*. Oxford: Blackwell.

Smith, A. (1776) *The Wealth of Nations*: *Books 1–111*. London: Penguin (1986 edition).

Smith, D.M. (2000a) 'Social justice revisited', *Environment and Planning A*, 32 (7): 1149–62.

Smith, D.M. (2000b) *Moral Geographies*. Edinburgh: Edinburgh University Press.

Smith, M. (1996) 'The empire filters back', *Urban Geography*, 17 (6): 502–24.

Smith, N. (1991) *Uneven Development*, 2nd edition. Oxford: Blackwell.

Smith, N. (1993) 'Homeless/global: scaling places', in J. Bird, B. Curtis, T. Putnam, G. Robertson and L. Tickner (eds), *Mapping the Futures: Local Cultures, Global Change*. London: Routledge, pp. 87–119.

SOPEMI (2000) *Trends in International Migration*. Paris: OECD.

Sparke, M. and Lawson, V. (2003) 'Entrepreneurial geographies', in J. Agnew, K. Mitchell and G. Toal (eds), *A Companion to Political Geography*. Oxford: Blackwell, pp. 315–34.

Staehli, I. (2003) 'Place', in J. Agnew, K. Mitchell and G. Toal (eds), *A Companion to Political Geography*. Oxford: Blackwell, pp. 158–70.

Stalker, P. (2000) *Workers Without Frontiers: The Impact of Globalisation on International Migration*. Boulder, CO: Lynne Reiner Publishers.

Storper, M. (1995) 'The resurgence of regional economies, ten years later', *European Urban and Regional Studies*, 2: 191–221.

Storper, M. and Walker, R. (1983) 'The theory of labour and the theory of location', *International Journal of Urban and Regional Research*, 7: 1–41.

Storper, M. and Walker, R. (1989) *The Capitalist Imperative*. Oxford: Blackwell.

Suárez, S. (2001) 'Political and economic motivations for labor control', *Studies in Comparative International Development*, 36 (2): 54–81.

Swyngedouw, E. (1989) 'The heart of a place', *Geografiska Annaler*, 71B: 31–42.

Swyngedouw, E. (1997) 'Neither global nor local', in K. Cox (ed.), *Spaces of Globalisation* New York: Guilford, pp. 137–66.

Swyngedouw, E. (2000a) 'Elite power, global forces, and the political economy of "glocal" development', in G. Clark, M. Feldman and M. Gertler (eds) *The Oxford Handbook of Economic Geography*. Oxford: Oxford University Press, pp. 541–58.

Swyngedouw, E. (2000b) 'The Marxian alternative', in E. Sheppard and T. Barnes (eds) *A Companion to Economic Geography*. Oxford: Blackwell, pp. 41–9.

Theodore, N. and Salmon, S. (1999) 'Globalizing capital, localizing labour?', *Space and Polity*, 3 (2): 153–69.

Thorpe, V. (1999) 'Global unionism: the challenge', in R. Munck and P. Waterman, pp. 218–29.

Thurow, L. (1975) *Generating Inequality*. New York: Basic Books.

Tickell, A. and Peck, J. (2003) 'Making global rules: globalization or neoliberalization?', in Peck, J. and Yeung, H. (eds) *Global Connections*. London: Sage, pp. 163–81.

Tilly, C. and Tilly, C. (1998) *Work Under Capitalism*. Boulder, CO: Westview Press.

Toynbee, P. (2003) *Hard Work*. London: Bloomsbury.

Treasury, H.M. (2001) *Productivity in the UK: The Regional Dimension*. London: H.M. Treasury.

Turnbull, P. (2000) 'Contesting globalization on the waterfront', *Politics and Society*, 28 (3): 367–91.

UNCTAD (2001) *The World Investment Report 2001*. New York: United Nations.

Van Hear, N. (1998) *New Diasporas*. London: UCL Press.

Vertovec, S. (1999) 'Conceiving and researching transnationalism', *Ethnic and Racial Studies*, 22 (2): 448–62.

Waddington, J. (ed.) (1999) *Globalization and Patterns of Labour Resistance*. London: Mansell.

Walsh, J. (2000) 'Organizing the scale of labor regulation in the USA', *Environment and Planning, A* 32: 1593–610.

Waterman, P. (1998) *Globalization, Social Movements and the New Internationalisms*. London: Mansell.

Waterman, P. (2001) 'Trade union internationalism in the age of Seattle', in P. Waterman and J. Wills, pp. 7–32.

Waterman, P. and Wills, J. (eds) (2001) *Space, Place and the New Labour Internationalisms*. Oxford: Blackwell.

Webb, S. and Webb, B. (1920) *Industrial Democracy*. London: Lawrence and Wishart.

Weiss, L. (1998) *The Myth of the Powerless State*. Cambridge: Polity Press.

Whatmore, S. and Thorne, L. (1997) 'Nourishing networks: alternative geographies of food ', in D. J. Goodman and M. J. Watts (eds), *Globalizing Food: Agrarian Questions and Global Restructuring*. London, Routledge, pp. 287–304.

Williams, C. and Windebank, J. (1998) *Informal Employment in the Advanced Economies*. London: Routledge.

Williams, R. (1975) *The Country and the City*. London: Paladin.

Willis, P. (1977) *Learning to Labour*. New York: Columbia University Press.

Wills, J. (1996) 'Geographies of trade unionism: translating traditions across space and time', *Antipode*, 28: 352–78.

Wills, J. (1998) 'Taking on the cosmocorps?', *Economic Geography*, 74: 111–30.

Wills, J. (2000) 'Being told and answering back', in J.R. Bryson, P.W. Daniels, N. Henry and J. Pollard (eds) *Knowledge, Space, Economy*. London: Routledge, pp. 261–76.

Wills, J. (2001) *Mapping Low Pay in East London. A Report Written for TELCO's Living Wage Campaign* (available from the author at http://www.qmw.ac.uk).

Wills, J. (2002) 'Community unionism and trade union renewal in the UK?' *Transactions of the Institute of British Geographers*, 26: 465–83.

Wills, J. and Simms, M. (2001) 'Building reciprocal community unionism in the UK', *Working Paper 4, Geographies of Organized Labour: the Reinvention of Trade Unionism in Millennial Britain ESRC Project*.

Womack, J.R., Jones, D.T. and Roos, D. (1990) *The Machine that Changed the World*. Rawson New York: Associates.

World Bank (2002) *World Development Indicators*. Washington: World Bank.

Yeung, H. W-C. (1994) 'Critical reviews of geographical perspectives on business organizations and the organization of production', *Progress in Human Geography*, 18: 460–90.

Zlotnick, H. (1998) 'International migration 1965–1996', *Population and Development Review*, 24 (3): 429–68.

Index

Lightning Source UK Ltd.
Milton Keynes UK
08 February 2010

149744UK00001B/27/P